Selected Titles in This Series

D1805569

731 **Jesús Bastero, Mario Milman, and Francisco J. Ruiz,** On the connection between weighted norm inequalities, commutators and real interpolation, 2001

730 **Suhyoung Choi,** The decomposition and classification of radiant affine 3-manifolds, 2001

729 **Michael Grosser, Eva Farkas, Michael Kunzinger, and Roland Steinbauer,** On the foundations of nonlinear generalized functions I and II, 2001

728 **Laura Smithies,** Equivariant analytic localization of group representations, 2001

727 **Anthony D. Blaom,** A geometric setting for Hamiltonian perturbation theory, 2001

726 **Victor L. Shapiro,** Singular quasilinearity and higher eigenvalues, 2001

725 **Jean-Pierre Rosay and Edgar Lee Stout,** Strong boundary values, analytic functionals, and nonlinear Paley-Wiener theory, 2001

724 **Lisa Carbone,** Non-uniform lattices on uniform trees, 2001

723 **Deborah M. King and John B. Strantzen,** Maximum entropy of cycles of even period, 2001

722 **Hernán Cendra, Jerrold E. Marsden, and Tudor S. Ratiu,** Lagrangian reduction by stages, 2001

721 **Ingrid C. Bauer,** Surfaces with $K^2 = 7$ and $p_g = 4$, 2001

720 **Palle E. T. Jorgensen,** Ruelle operators: Functions which are harmonic with respect to a transfer operator, 2001

719 **Steve Hofmann and John L. Lewis,** The Dirichlet problem for parabolic operators with singular drift terms, 2001

718 **Bernhard Lani-Wayda,** Wandering solutions of delay equations with sine-like feedback, 2001

717 **Ron Brown,** Frobenius groups and classical maximal orders, 2001

716 **John H. Palmieri,** Stable homotopy over the Steenrod algebra, 2001

715 **W. N. Everitt and L. Markus,** Multi-interval linear ordinary boundary value problems and complex symplectic algebra, 2001

714 **Earl Berkson, Jean Bourgain, and Aleksander Pełczynski,** Canonical Sobolev projections of weak type $(1, 1)$, 2001

713 **Dorina Mitrea, Marius Mitrea, and Michael Taylor,** Layer potentials, the Hodge Laplacian, and global boundary problems in nonsmooth Riemannian manifolds, 2001

712 **Raúl E. Curto and Woo Young Lee,** Joint hyponormality of Toeplitz pairs, 2001

711 **V. G. Kac, C. Martinez, and E. Zelmanov,** Graded simple Jordan superalgebras of growth one, 2001

710 **Brian Marcus and Selim Tuncel,** Resolving Markov chains onto Bernoulli shifts via positive polynomials, 2001

709 **B. V. Rajarama Bhat,** Cocylces of CCR flows, 2001

708 **William M. Kantor and Ákos Seress,** Black box classical groups, 2001

707 **Henning Krause,** The spectrum of a module category, 2001

706 **Jonathan Brundan, Richard Dipper, and Alexander Kleshchev,** Quantum Linear groups and representations of $GL_n(\mathbb{F}_q)$, 2001

705 **I. Moerdijk and J. J. C. Vermeulen,** Proper maps of toposes, 2000

704 **Jeff Hooper, Victor Snaith, and Min van Tran,** The second Chinburg conjecture for quaternion fields, 2000

703 **Erik Guentner, Nigel Higson, and Jody Trout,** Equivariant E-theory for C^*-algebras, 2000

702 **Ilijas Farah,** Analytic guotients: Theory of liftings for quotients over analytic ideals on the integers, 2000

701 **Paul Selick and Jie Wu,** On natural coalgebra decompositions of tensor algebras and loop suspensions, 2000

(Continued in the back of this publication)

On the Connection between Weighted Norm Inequalities, Commutators and Real Interpolation

MEMOIRS
of the
American Mathematical Society

Number 731

On the Connection between Weighted Norm Inequalities, Commutators and Real Interpolation

Jesús Bastero
Mario Milman
Francisco J. Ruiz

November 2001 • Volume 154 • Number 731 (second of 5 numbers) • ISSN 0065-9266

American Mathematical Society
Providence, Rhode Island

2000 *Mathematics Subject Classification.*
Primary 46E30, 46B70.

Library of Congress Cataloging-in-Publication Data

Bastero, J. (Jesús), 1950–
On the connection between weighted inequalities, commutators, and real interpolation / Jesús Bastero, Mario Milman, Francisco J. Ruiz.
p. cm. — (Memoirs of the American Mathematical Society, ISSN 0065-9266 ; no. 731)
"Volume 154, number 731 (second of 5 numbers)."
Includes bibliographical references.
ISBN 0-8218-2734-0 (alk. paper)
1. Fourier analysis. 2. Interpolation spaces. 3. Function spaces. I. Milman, Mario. II. Ruiz, Francisco J., 1956– III. Title. IV. Series.

QA3 .A57 no. 731
[QA403.5]
510 s—dc21
[515′.2433] 2001034316

Memoirs of the American Mathematical Society

This journal is devoted entirely to research in pure and applied mathematics.

Subscription information. The 2001 subscription begins with volume 149 and consists of six mailings, each containing one or more numbers. Subscription prices for 2001 are \$494 list, \$395 institutional member. A late charge of 10% of the subscription price will be imposed on orders received from nonmembers after January 1 of the subscription year. Subscribers outside the United States and India must pay a postage surcharge of \$31; subscribers in India must pay a postage surcharge of \$43. Expedited delivery to destinations in North America \$35; elsewhere \$130. Each number may be ordered separately; *please specify number* when ordering an individual number. For prices and titles of recently released numbers, see the New Publications sections of the *Notices of the American Mathematical Society.*

Back number information. For back issues see the *AMS Catalog of Publications.*

Subscriptions and orders should be addressed to the American Mathematical Society, P. O. Box 845904, Boston, MA 02284-5904. *All orders must be accompanied by payment.* Other correspondence should be addressed to Box 6248, Providence, RI 02940-6248.

Memoirs of the American Mathematical Society is published bimonthly (each volume consisting usually of more than one number) by the American Mathematical Society at 201 Charles Street, Providence, RI 02904-2294. Periodicals postage paid at Providence, RI. Postmaster: Send address changes to Memoirs, American Mathematical Society, P. O. Box 6248, Providence, RI 02940-6248.

This publication is indexed in *Science Citation Index®, SciSearch®, Research Alert®, CompuMath Citation Index®, Current Contents®/Physical, Chemical & Earth Sciences.*
Printed in the United States of America.

♾ The paper used in this book is acid-free and falls within the guidelines established to ensure permanence and durability.
Visit the AMS home page at URL: `http://www.ams.org/`

10 9 8 7 6 5 4 3 2 1 06 05 04 03 02 01

Contents

Abstract

We show that the class of weights w for which the Calderón operator is bounded on $L^p(w)$ can be used to develop a theory of real interpolation which is more general and exhibits new features when compared to the usual variants of the Lions-Peetre methods. In particular we obtain extrapolation theorems (in the sense of Rubio de Francia's theory) and reiteration theorems for these methods. We also consider interpolation methods associated with the classes of weights for which the Calderón operator is bounded on weighted Lorentz spaces and obtain similar results. We extend the commutator theorems associated with the real method of interpolation in several directions. We obtain weighted norm inequalities for higher order commutators as well as commutators of fractional order. One application of our results gives new weighted norm inequalities for higher order commutators of singular integrals with multiplications by BMO functions. We also introduce analogs of the space BMO in order to consider the relationship between commutators for Calderón type operators and their corresponding classes of weights.

2000 *Mathematics Subject Classification.* Primary 46B70, 46E30. Secondary 42B25.

Key words and phrases. Calderón operator, weighted norm inequalities, interpolation of operators, commutators, BMO.

1. Introduction

Many of the problems we study in analysis can be rephrased in terms of the study of the action of operators in function spaces. The more we understand about the action of a given operator in as many as possible function spaces, the more we understand about the nature of the problem under consideration. Interpolation is a very useful tool for this purpose as it provides us with methods to obtain new estimates from old ones.

In recent years a theory of weighted norm inequalities for classical operators has been developed. It has led to a deeper understanding of many important problems in analysis. Moreover, it has led to the discovery of unexpected relationships between different areas of analysis.

At the basis of these developments it is the celebrated theory of Muckenhoupt for the maximal operator of Hardy-Littlewood M (see [**Mu1**]). Let $1 < p < \infty$, then we have $M : L^p(w) \to L^p(w)$ if and only if the weight w belongs to the class $\mathcal{A}_p$. The $\mathcal{A}_p$ classes of weights admit a concrete description and their properties have been intensively investigated. Muckenhoupt's results led to an extensive study of weighted norm inequalities for classical operators (singular integral, multipliers, square functions, ...). This has also uncovered deep connections between classes of weights and function spaces, like $\mathcal{A}_p$ and BMO.

A beautiful connection between the theory of weighted norm inequalities and the theory of factorization of operators on Banach spaces is given by Rubio de Francia's extrapolation theorem. A simple version of this result states that if T is a bounded linear operator on $L^2(w)$, for all $w \in \mathcal{A}_2$, then T is also bounded on $L^p(w)$, for all $1 < p < \infty$ and for all $w \in \mathcal{A}_p$.

There is a close connection between interpolation theory and weighted norm inequalities. In particular, interpolation theory provides methods to obtain rearrangement inequalities for operators to which one can then apply weighted norm inequalities. While interpolation has been useful in the study of weighted norm inequalities it seems to us that a deeper study of the connection between interpolation theory and weighted norm inequalities is still to be developed.

For example, what would be the analog of Rubio's theorem for interpolation scales?, what would be the natural classes of weights that we should consider in a general theory of real interpolation scales?

In the first part of this paper we consider these questions and establish an extrapolation theorem for operators acting on weighted real interpolation spaces. In order to even formulate these results we are forced to generalize the classical theory of Lions-Peetre through the consideration of weighted L^p spaces. In this

---

Received by the editor November 20, 1996; and in revised form December 28, 1999.
The first author was partially supported by DGES.
The third author was partially supported by DGES.

setting some of the characteristic features of real interpolation are blended with the theory of weights.

In the second part of the paper we study the connection of the theory of weighted interpolation developed in the first half with the so called commutator estimates arising from the real method of interpolation. Here the motivation for our work comes from recent striking work by Muller [**Mu**] and Coifman, Lions, Meyer, and Semmes [**CLMS**] and the rapidly growing body of literature generated through their influence. Indeed, these papers have generated considerable new impetus for the applications of real variable techniques to estimate the size of certain nonlinear expressions that appear in nonlinear PDE's. These operations (e.g. Jacobians, Null Lagrangians, etc) can be estimated because of subtle cancellations. For a sample of recent results we refer to [**EM**], [**Li**], [**LZ**], [**LMZ**], [**Mi1**], [**Se**] and the references quoted therein. The higher integrability of these nonlinear operators is crucial to establish their compactness in suitable weak topologies. These developments provide a new framework to study questions of the theory of compensated compactness developed by Murat and Tartar (cf. [**CLMS**] and the references therein). Many of these estimates can be established using commutator theorems for singular integrals and methods arising from interpolation theory (cf. [**RW**], [**JRW**], [**IS**], [**Mi**]). These commutator estimates are also important in the regularity theory of quasilinear second order elliptic equations under minimal assumptions on the coefficients (cf. the survey [**Ch**] and the references therein).

Interpolation theory plays an important role in these studies. On the one hand it can be used to establish many of these concrete estimates while on the other it provides tools to study higher integrability of non linear operations, due to cancellations, in a very general setting. In fact in this generalized setting many of the arguments are simpler and the role of the cancellations apparent. Moreover, interpolation methods also point to expressions that exhibit higher order cancellations and which therefore are bounded in better spaces than one could have predicted from size considerations only. The theory, originally started in [**RW**] and [**JRW**], has been extended and applied in several directions. We refer to [**Mi**] and [**MR**] for an extended discussion, as well as a detailed list of contributions. Apart from its applications to Elasticity Theory, Harmonic Analysis, and Partial Differential Equations the theory has also been applied in other areas of functional analysis. In particular we refer to [**Ka**], [**Ka1**], [**Ka2**] and the review papers quoted above for more references.

It therefore seemed to us of timely interest to extend the general theory of commutators associated with the real method of interpolation. In our development in this paper we focus on two directions. On the one hand we treat commutators of bounded operators with a general class of nonlinear operators including higher order commutators of fractional order. On the other we also consider weighted norm inequalities for these operators in a very general context. Our methods are based, and generalize, the analysis given in [**Mi**] and [**Mi3**].

In order to explain in some more detail what we do let us now recall some basic facts and definitions.

If f is a measurable function defined on the interval $(0, \infty)$, the Calderón operator S is defined by

$$\begin{aligned} Sf(t) &= \int_0^\infty \min\{1/x, 1/t\} f(x)dx \\ &= \frac{1}{t}\int_0^t f(x)dx + \int_t^\infty \frac{f(x)}{x}dx \\ &= P(f)(t) + Q(f)(t). \end{aligned}$$

P is called the Hardy operator and Q is its adjoint. The Calderón operator plays an important role in interpolation theory. In particular it controls the relationship between the K-method and the J-method of interpolation, a result due to Brundyi and Krugljak (see [**BK**]). Indeed, given a compatible couple of Banach spaces $\bar{A} = (A_0, A_1)$, if $a \in A_0 + A_1$ is an element for which there exists a representation $a = \int_0^\infty \frac{a(t)}{t}dt$, with $a(t) \in A_0 \cap A_1$ then

$$\frac{K(t, a; \bar{A})}{t} \leq S\left(\frac{J(x, a(x); \bar{A})}{x}\right)(t),$$

where $K(t, a; \bar{A})$ is the classical K-functional of interpolation, i.e.

$$K(t, a; \bar{A}) = \inf\{\|a_0\|_{A_0} + t\|a_1\|_{A_1}\},$$

where the inf runs over all possible decompositions $a = a_0 + a_1$ with $a_i \in A_i$, and J is the classical J-functional given by

$$J(t, a; \bar{A}) = \max\{\|a\|_{A_0}, t\|a\|_{A_1}\}$$

for the elements $a \in A_0 \cap A_1$.

In the first part of the paper we show that the class of weights that controls the weighted L^p estimates for the Calderón operator can be used develop a rich theory of interpolation which includes some novel features that are not present in the classical spaces of Lions-Peetre.

Heuristically the Calderón operator is 'the only' operator we should consider in order to obtain a universal class of weights that is suitable for interpolation theory, since it majorizes in a suitable sense all other operations in a given interpolation segment. This should be compared with the corresponding theory of $\mathcal{A}_p$ Muckenhoupt weights, based on the maximal operator of Hardy and Littlewood.

Let w be a weight on $(0, \infty)$, i.e., w is a measurable function, $w > 0$ a.e. with respect to the Lebesgue measure. We denote by $L^p(w)$, $1 \leq p < \infty$, the classes of Lebesgue measurable functions f defined on the interval $(0, \infty)$ such that

$$\|f\|_{p,w} = \left(\int_0^\infty |f(t)|^p w(t)dt\right)^{1/p} < +\infty.$$

For $p = \infty$ the corresponding space, $L^\infty(w)$, is defined using the norm

$$\|f\|_{\infty,w} = \|fw\|_\infty < +\infty.$$

In [**GC**] these spaces are denoted by $w^{-1}L^\infty$ and we shall also adopt this notation, when convenient, in what follows.

Let w be a weight and let $1 \leq p \leq \infty$. We define $\bar{A}_{w,p;K}$ as the class of vectors $a \in A_0 + A_1$ for which the function $t^{-1}K(t,a;\bar{A}) \in L^p(w)$. For $a \in A_{w,p;K}$ we let

$$\|a\|_{\bar{A}_{w,p;K}} = \left(\int_0^\infty \left(\frac{K(t,a;\bar{A})}{t}\right)^p w(t)dt\right)^{1/p}.$$

If we consider the J-method of interpolation, we define $\bar{A}_{w,p;J}$ as the class of elements $a \in A_0 + A_1$ for which there exists a representation $a = \int_0^\infty \frac{a(t)}{t}dt$ with $a(t) \in A_0 \cap A_1$ satisfying $t^{-1}J(t,a(t);\bar{A}) \in L^p(w)$. For this class we consider the corresponding norms

$$\|a\|_{\bar{A}_{w,p;J}} = \inf \left(\int_0^\infty \left(\frac{J(t,a(t);\bar{A})}{t}\right)^p w(t)dt\right)^{1/p}$$

where the inf runs over all possible representations of a.

The classical scales of real interpolation spaces of Lions-Peetre correspond to the power weights $w = t^{p-p\theta-1}$. Apart from this case, the most studied classes of interpolation spaces are the so called "functional parameter" and the "quasipower" cases. These spaces are defined by a slightly more general class of weights than powers. If v is a quasipower weight, the interpolation scales associated with the functional parameter are classically defined by $(A_0, A_1)_{v,p;K} = \{a \in A_0 + A_1; \|a\|_{v,p;K} < \infty\}$ where

$$\|a\|_{v,p;K} = \int_0^\infty \left(K(t,a;\bar{A})v\right)^p \frac{dt}{t}.$$

The study of interpolation spaces defined using these classes of weights was initiated in [**K**] and continued in [**G**] and many other papers, cf. [**BK**]. In our notation we have

$$(A_0, A_1)_{v,p;K} = \bar{A}_{w,p;K}$$

with $w = v^p t^{p-1}$.

The point of view advocated in this paper is related to work by Sagher [**Sg**], where Calderón type of weights are used to extend classical interpolation theorems and also with a question raised by E. Hernández and J. Soria. In [**HS**] the authors showed that weights of the form $w = v^p t^{p-1}$, v a quasipower weight, are in the class $\mathcal{C}_p$ (see Theorem 4.1 in [**HS**]) and asked for a general theory of real interpolation method with weights in $\mathcal{C}_p$, which "would be more general than the existing ones."

In this paper we have developed such a theory in detail. Furthermore we have proved that for $p = 1$ the scales introduced by this method can be represented using quasipower weights (see Proposition 3.8 below). More generally, for $p > 1$, we indicate the relationship between our interpolation scales with those derived using the classical functional parameter approach (see Proposition 3.9 below).

One can also consider other classes of weights and their corresponding associated interpolation methods. In this paper we also develop, in some detail, the theory associated with the $\mathcal{B}_p$ classes introduced in [**AM**] in the study of the Hardy operator P acting on weighted $L^p(w)$ spaces but restricted to non-increasing functions. Their direct import in interpolation theory can be seen from the fact that, if $A_0 \cap A_1$ is dense in A_0, we have (cf. [**BS**])

$$\frac{1}{t}K(t,a;\bar{A}) = \frac{1}{t}\int_0^t k(x,a,\bar{A})dx = P(k(\cdot,a;\bar{A}))(t),$$

where k is the derivative of the functional K. Note that both $\frac{1}{t}K(t,a;\bar{A})$ and $k(t,a,\bar{A})$ are decreasing functions. Therefore the $\mathcal{B}_p$ weights control the equivalence between the K and k methods of interpolation (cf. §4 below).

Let us point out that it is only in this generalized setting that we can formulate an analogue of Rubio de Francia's extrapolation theorem. To see better the role that generalized weights play let us recall that by reiteration,

$$(A_{\theta_0,p_0;K}, A_{\theta_1,p_1;K})_{\theta,q} = (A_0, A_1)_{\eta,q;K},$$

where $\eta = (1-\theta)\theta_0 + \theta\theta_1$. Thus, the second index is not important for reiteration. Consequently for power weights, that is in the classical setting of the Lions-Peetre spaces, reiteration plays the role of extrapolation in the sense of Rubio de Francia and thus it is not of interest in this context. However, in the setting of the generalized $\mathcal{C}_p$ -weights, the following extrapolation result holds (see §3 Theorem 3.10 below.)

THEOREM. *Let $\bar{A}$, $\bar{B}$ two compatible pairs of Banach spaces, and let T be a linear operator bounded from $\bar{A}_{w,p;K}$ into $\bar{B}_{w,p;K}$ for some p, $1 \le p < \infty$, and for all $w \in \mathcal{C}_p$, with norm that depends only upon the $\mathcal{C}_p$-constant for w. Then T is also bounded from $\bar{A}_{v,q;K}$ into $\bar{B}_{v,q;K}$ for any q, $1 < q < \infty$, and for all $v \in \mathcal{C}_q$ with norm that depends only upon the $\mathcal{C}_q$-constant for v.*

As an important corollary of our presentation, we obtain a new result even for classical functional parameter in terms of extrapolation (see Corollary 3.11).

We shall also show in §5 that if we have a family of estimates for a bounded operator on the classical Lions-Peetre scale and if these estimates do not 'blow up' too fast then an extension of the previous extrapolation theorem holds and can be obtained using the extrapolation method of Jawerth and Milman (cf. [**JM**]). This provides us with a simple and effective method to prove weighted Lorentz estimates of the type considered by Ariño and Muckenhoupt (cf. [**AM**]).

Real interpolation spaces are constructed by means of decomposing effectively their elements. Given a bounded operator T between two Banach pairs $\bar{A}$ and $\bar{B}$, these decompositions can be applied before and after applying T. These considerations lead to the construction of operators that are based on these optimal decompositions, we shall call them $\Omega_{K,\bar{A}}$ and $\Omega_{K,\bar{B}}$ and lead to the study of the commutator

$$[T,\Omega_K]a = (T\Omega_{K,\bar{A}} - \Omega_{K,\bar{B}}T)a,$$

where $a \in \bar{A}_{w,p;K}$ (cf. [**RW**], [**JRW**]).

For specific pairs these commutators are some of the classical commutators studied in classical analysis. In the second part of the paper we consider the role that weights play in the theory. In particular we consider commutators with operators Ω which are constructed using these weights, as well as norm estimates of these operations.

One concrete application of our results provides new weighted norm inequalities for commutators of singular integrals with multiplications with BMO functions. Moreover, as indicated above, these estimates, when combined with the results described in [**Ch**], provide new applications to the regularity theory of quasilinear elliptic equations.

We also remark that our methods also apply to other operators including fractional integrals, and given the generality of our assumptions, our methods also give analogous estimates for operators acting on spaces of homogeneous type. Our results also give weighted norm inequalities for commutators of other classical operators of Harmonic Analysis. For background on these developments in the setting of Lorentz spaces we refer to [**Sw**]. It is relevant to mention here the recent work [**Pe**] on weighted norm inequalities for commutators which complements our work here. The direct connection between the two subjects treated in this paper is developed in the last part of the paper. Recall that weighted norm inequalities for singular integral operators, commutator estimates for singular integrals with multiplication by BMO functions are closely related to the connection between $\mathcal{A}_p$ weights and BMO. In our context these considerations also lead us to consider variants of the space BMO in the context of the weights $\mathcal{C}_p$. Thus, in analogy with the classical theory of $\mathcal{A}_p$ weights, the set of logarithms of $\mathcal{B}_p$ weights gives BMO type spaces. We also construct its predual, via a suitable atomic theory.

Applications play an important role in our development. Throughout the paper we consider several applications and examples relating our results to singular integrals, multipliers, H^p spaces, Tent spaces, Hardy-Sobolev spaces, approximation theory, Schatten ideals, Dirichlet spaces, etc. Concrete new estimates (weighted and unweighted) for commutators are given in different contexts including estimates for Jacobians of maps and other operations with sufficient cancellations.

We shall now review the organization of the paper. In §2 we study the Calderón weights $\mathcal{C}_p$ and compare them with Kalugina weights and quasipower weights. In §3 we study the real interpolation spaces associated with Calderón weights, paying special attention to reiteration and extrapolation results. In §4 we briefly consider interpolation methods associated with $\mathcal{B}_p$ weights, in §5 we consider a connection between Rubio de Francia's theory and the theory of extrapolation of Jawerth-Milman, in §6 we discuss some applications to the study of other scales of function spaces including Lorentz spaces, weighted Tent spaces, ideals of operators, Hardy spaces, Hardy-Sobolev spaces, Dirichlet spaces, in §7 we develop our theory of commutators, in §8 we discuss generalized commutators including those of fractional order, in §9 we extend our results to the setting of Quasi-Banach spaces. In §10 we discuss some applications of our results to singular integrals and to results related with compensated compactness, in §11 we introduce the space BMO and H^1 type of spaces associated to the weights under study and show in §12 duality and interpolation results for these spaces as well as new applications illustrating the connection between the two parts of the paper.

In conclusion we should mention that there also several commutator theorems associated with the complex method of interpolation and the interested reader should consult [**RW**], [**CCMS**], [**R**], as well as the unified theories developed in [**CCS**] and [**CKMR**] and their references.

Throughout the paper we shall follow the notation and terminology of [**BL**].

ACKNOWLEDGMENT. *We would like to thank the referee for her/his comments and for helpful suggestions to improve the presentation of the paper.*

2. Calderón weights

Let P, Q be the operators defined in the Introduction. Results by Muckenhoupt (see [**Mu2**], [**Ma**]), which extend Hardy's inequalities, ensure that:

A. *$Pf \in L^p(w)$ for all $f \in L^p(w)$ $(1 \leq p < \infty)$ if and only if there exists a constant $C > 0$ such that for almost all $t > 0$*

$$\left(\int_t^\infty \frac{w(x)}{x^p} dx\right)^{1/p} \left(\int_0^t w(x)^{-p'/p} dx\right)^{1/p'} \leq C \tag{M_p}$$

for $1 < p < \infty$, or

$$\int_t^\infty \frac{w(x)}{x} dx \leq C w(t) \tag{M_1}$$

for $p = 1$,

and

B. *$Qf \in L^p(w)$ for all $f \in L^p(w)$ $(1 \leq p < \infty)$ if and only if there exists a constant $C > 0$ such that for almost all $t > 0$*

$$\left(\int_0^t w(x) dx\right)^{1/p} \left(\int_t^\infty \frac{w(x)^{-p'/p}}{x^{p'}} dx\right)^{1/p'} \leq C \tag{M^p}$$

for $1 < p < \infty$, or

$$\frac{1}{t} \int_0^t w(x) dx \leq C w(t) \tag{M^1}$$

for $p = 1$.

With the exception of the trivial case $w = 0$ a.e., the conditions above imply that $w > 0$ a.e. and all the integrals appearing in (M_p) and (M^p) are finite.

It will be important for us to relate the weighted norm inequalities for the Calderón or Hardy operators in terms of the conditions on the weights. It follows from [**Mu2**] that, for $1 \leq p < \infty$,

$$\|w\|_{M_p} \leq \|P\|_{L^p(w) \to L^p(w)} \leq p^{1/p} p'^{1/p'} \|w\|_{M_p} \leq 4\|w\|_{M_p}, \tag{2.1}$$

where $\|w\|_{M_p}$ is the infimum of the constants C appearing in the definition of the M_p condition. Likewise

$$\|w\|_{M^p} \leq \|Q\|_{L^p(w) \to L^p(w)} \leq p^{1/p} p'^{1/p'} \|w\|_{M^p} \leq 4\|w\|_{M^p} \tag{2.2}$$

where $\|w\|_{M^p}$ is the infimum of the constants C appearing in the definition of the M^p condition.

Note that

$$1 \leq \|w\|_{M^p} \|w\|_{M_p}. \tag{2.3}$$

Indeed,

$$1=\int_0^t dx\int_t^\infty \frac{dx}{x^2}$$

$$\leq\left(\int_0^t w\right)^{1/p}\left(\int_0^t w^{-p'/p}\right)^{1/p'}\left(\int_t^\infty \frac{w}{x^p}dx\right)^{1/p}\left(\int_t^\infty \frac{w^{-p'/p}}{x^{p'}}dx\right)^{1/p'}.$$

DEFINITION 2.1. *Let $1\leq p<\infty$. We say that a weight $w\in\mathcal{C}_p$ if it satisfies the conditions M_p and M^p simultaneously. We say that a weight $w\in\mathcal{C}_\infty$ if $w^{-1}\in\mathcal{C}_1$.*

We also define the corresponding "norms" for the $\mathcal{C}_p$ weights $(1\leq p<\infty)$ by

$$\|w\|_{\mathcal{C}_p}=\|w\|_{M^p}+\|w\|_{M_p}. \tag{2.4}$$

Since the Calderón operator is $S=P+Q$, and P, Q are positive, the class $\mathcal{C}_p$ is actually the class of weights for which S is bounded from $L^p(w)$ into $L^p(w)$. Then

$$\frac{1}{2}\|w\|_{\mathcal{C}_p}\leq\|S\|_{L^p(w)\to L^p(w)}\leq 4\|w\|_{\mathcal{C}_p}. \tag{2.5}$$

For $p=\infty$ it is easy to see that S is bounded from $L^\infty(w)$ into $L^\infty(w)$ if and only if $S(w^{-1})(x)\leq Cw^{-1}(x)$, a.e. x, for some constant $C>0$, that is, $w^{-1}\in\mathcal{C}_1$.

The operator S is positive and selfadjoint, and since the class $\mathcal{C}_1$ coincides with the class of weights for which there exists a constant $C>0$ such that $Sw\leq Cw$, we can apply to these classes the strong machinery developed by Rubio de Francia and prove for them similar properties to the ones satisfied by the classical $\mathcal{A}_p$ classes, i.e., factorization and extrapolation. We also study reverse Hölder type inequalities for them.

PROPOSITION 2.2. *Let $1\leq p<\infty$. A weight $w\in\mathcal{C}_p$ if and only if there exist two weights w_0, $w_1\in\mathcal{C}_1$ such that $w=w_0w_1^{1-p}$.*

PROOF. If $w\in\mathcal{C}_p$ then the desired factorization can be obtained using Rubio de Francia's factorization theorem (cf. [**GR**]). Indeed, we just need to recall that $S=P+Q$ is selfadjoint and the following duality property holds: $w\in\mathcal{C}_p$ if and only if $w^{-p'/p}\in\mathcal{C}_{p'}$. Let us also remark that [**Mu2**] has constructive factorizations results of the same type for each of the classes M_p and M^p of weights.

The converse can be obtained by direct computation. Suppose that w_i, $i=0,1$, are two $\mathcal{C}_1$-weights. Let $w=w_0w_1^{1-p}$. By definition we can assume that

$$t^{-1}\int_0^t w_i(x)dx+\int_t^\infty\left(\frac{t}{x}\right)w_i(x)dx\leq Cw_i(t) \tag{2.6}$$

for all $t>0$, $i=0,1$, and for some constant $C>0$. We will now prove that condition (M_p) holds for w.

$$I=\left(\int_t^\infty\frac{w(x)}{x^p}dx\right)^{1/p}\left(\int_0^t w(x)^{-p'/p}dx\right)^{1/p'}$$

$$=\left(\int_t^\infty\frac{w_0(x)w_1(x)^{1-p}}{x^p}dx\right)^{1/p}\left(\int_0^t w_0(x)^{-p'/p}w_1(x)dx\right)^{1/p'}$$

which, by (2.6) is bounded by a constant times

$$\left(\int_t^\infty \frac{w_0(x)}{x^p}\left(\frac{1}{x}\int_0^x w_1\right)^{1-p}dx\right)^{1/p}\left(\int_0^t w_1(x)\left(\int_x^\infty \frac{w_0(s)}{s}ds\right)^{-p'/p}dx\right)^{1/p'}.$$

Since $t < x$ implies

$$\int_0^x w_1 \geq \int_0^t w_1$$

and $t > x$ implies

$$\int_x^\infty s^{-1}w_0(s)ds \leq \int_t^\infty s^{-1}w_0(s)ds,$$

then

$$I \leq C\left(\int_t^\infty \frac{w_0}{x}dx\right)^{1/p}\left(\int_0^t w_1\right)^{-1/p'}\left(\int_0^t w_1\right)^{1/p'}\left(\int_t^\infty \frac{w_0}{x}dx\right)^{-1/p} \leq C.$$

Since the verification of the condition (M_p) can be obtained in a similar fashion we shall skip the details. □

REMARK. The converse of Proposition (2.2) is also a consequence of the main result in [**Bl**], since S is an operator defined by means of a positive kernel. Another direct consequence of the result in [**Bl**] is the following factorization for M_p and M^p weights:

PROPOSITION 2.2'. *Let $1 \leq p < \infty$. A weight $w \in M_p$ (resp. M^p) if and only if there exist two weights $w_0 \in M_1$ (resp. M^1), $w_1 \in M^1$ (resp. M_1) such that $w = w_0 w_1^{1-p}$.*

The following result can be seen as a type of reverse Hölder inequality for $\mathcal{C}_1$ weights. Results of this type have applications in interpolation theory for rearrangement invariant Banach spaces that we shall pursue elsewhere (see [**BR**], [**BMR**]).

PROPOSITION 2.3. *If $w \in \mathcal{C}_1$ then there exists $\epsilon > 0$ such that, for some constant $C > 0$, w satisfies*

$$\int_0^t \left(\frac{t}{x}\right)^\epsilon w(x)dx + \int_t^\infty \left(\frac{t}{x}\right)^{1-\epsilon} w(x)dx \leq Ctw(t)$$

for all $t > 0$.

PROOF. Let C be the constant such that $P(w)(t) + Q(w)(t) \leq Cw(t)$, $\forall t > 0$. In order to prove the proposition we shall use the following facts which can be easily proved by Fubini's theorem and by induction

$$P \circ Q = Q \circ P \tag{2.7}$$

$$P \circ Q^{(k)}(w) \leq C^k w, \text{ for all } k \geq 1 \tag{2.8}$$

$$Q \circ P^{(k)}(w) \leq p^{-1}C^k w, \text{ for all } k \geq 1 \tag{2.9}$$

$$Q^{(k)}(w)(s) = \int_s^\infty \frac{w(x)}{x} \frac{\log(x/s)^{k-1}}{(k-1)!} dx \tag{2.10}$$

$$P^{(k)}(w)(s) = \frac{1}{s} \int_0^s w(x) \frac{\log(s/x)^{k-1}}{(k-1)!} dx, \tag{2.11}$$

where $P^{(k)} = P \circ \overset{(k}{\ldots} \circ P$ and $Q^{(k)} = Q \circ \overset{(k}{\ldots} \circ Q$.

Now, in order to prove proposition 2.3, pick a constant $\epsilon > 0$ such that $\epsilon C < 1$. By (2.10) and (2.11), we have

$$\sum_{k=1}^\infty \epsilon^{k-1} Q^{(k)}(w)(s) = s^{-\epsilon} \int_s^\infty \frac{w(x)}{x^{1-\epsilon}} dx$$

and

$$\sum_{k=1}^\infty \epsilon^{k-1} P^{(k)}(w)(s) = \frac{1}{s^{1-\epsilon}} \int_0^s \frac{w(x)}{x^\epsilon} dx.$$

Moreover, by (2.8) and (2.9)

$$P \circ \left(\sum_{k=1}^\infty \epsilon^{k-1} Q^{(k)} \right) (w) \leq C' w$$

and

$$Q \circ \left(\sum_{k=1}^\infty \epsilon^{k-1} P^{k} \right) (w) \leq C' w.$$

This means that

$$\frac{1}{t} \int_0^t \frac{w(x)}{x^{1-\epsilon}} dx \int_0^x s^{-\epsilon} ds + \frac{1}{t} \int_t^\infty \frac{w(x)}{x^{1-\epsilon}} dx \int_0^t s^{-\epsilon} ds =$$
$$= \frac{1}{t} \int_0^t w(x) dx + t^{-\epsilon} \int_t^\infty \frac{w(x)}{x^{1-\epsilon}} dx \leq C' w(t),$$

and

$$\int_0^t \frac{w(x)}{x^\epsilon} dx \int_t^\infty \frac{ds}{s^{2-\epsilon}} + \int_t^\infty \frac{w(x)}{x^\epsilon} dx \int_x^\infty \frac{ds}{s^{2-\epsilon}} =$$
$$= \frac{1}{t} \int_0^t \left(\frac{t}{x} \right)^\epsilon w(x) dx + \int_t^\infty \frac{w(x)}{x} dx \leq C' w(t)$$

as we wished to show. □

REMARKS.

i) If a weight satisfies the corresponding inequality for some $\epsilon > 0$ it also does for any $0 < \delta < \epsilon$.

ii) The method of proof of Proposition 2.3 also applies to the $\mathcal{B}_p$ weights (see Definition 4.1 below) studied by Ariño and Muckenhoupt (cf. [**AM**]). For these weights it is known that $w \in \mathcal{B}_p \Rightarrow w \in \mathcal{B}_{p-\epsilon}$, for some $\epsilon > 0$ (see [**AM**], [**N1**]). We can obtain a simple direct proof of this fact using arguments similar to the ones in Proposition 2.3 (cf. [**BR**]).

As a consequence of propositions 2.2 and 2.3 we achieve a kind of reverse Hölder type inequality for the class $\mathcal{C}_p$.

PROPOSITION 2.4. *If $w \in \mathcal{C}_p$ then there exists $\epsilon > 0$ such that $x^{-\epsilon p} w(x) \in M^p$ and $x^{\epsilon p} w(x) \in M_p$.*

PROOF. According to proposition 2.2, $w = w_0 w_1^{1-p}$ and there exist constants $C > 0$ and $\epsilon > 0$ such that

$$t^{-1} \int_0^t \frac{w_i(x)}{x^\epsilon} dx + \int_t^\infty \left(\frac{t}{x}\right)^{1-2\epsilon} \frac{w_i(x)}{x^\epsilon} dx \le C \frac{w_i(t)}{t^\epsilon}$$

for all $t > 0$, $i = 0, 1$. Hence, by an argument similar to the one in the converse part of the proof of proposition 2.2 we see that the weight $x^{-\epsilon p} w(x) \in M^p$ and the weight $x^{\epsilon p} w(x) \in M_p$. □

REMARK. The last proposition ensures that a weight w in the class $\mathcal{C}_p$ supports stronger integrability conditions at 0 and at ∞.

We are now going to study analogs of Rubio de Francia's extrapolation theorem in the context of $\mathcal{C}_p$ weights.

First, we need the following convexity property.

LEMMA 2.5. *Let $1 < p < \infty$. If $u \in \mathcal{C}_1$, $v \in \mathcal{C}_p$ and $0 \le s \le 1$ then $u^s v^{1-s} \in \mathcal{C}_{s+p(1-s)}$.*

PROOF. In order to check the condition $M_{s+p(1-s)}$ we use Hölder's inequality with exponents $1/s$ and $1/(1-s)$ in the first factor, and in the second one the fact that if $u \in \mathcal{C}_1$ and $\beta < 0$ then for $x < t$,

$$u(x)^\beta \le C \left(\int_t^\infty \frac{u(y)}{y} dy \right)^\beta .$$

Therefore, we have

$$\left(\int_t^\infty \frac{u(x)^s v(x)^{1-s}}{x^{s+p(1-s)}} dx \right) \left(\int_0^t (u(x)^s v(x)^{1-s})^{-\frac{1}{(1-s)(p-1)}} dx \right)^{(1-s)(p-1)} \le$$

$$C \left(\int_t^\infty \frac{u(x)}{x} dx \right)^s \left(\int_t^\infty \frac{v(x)}{x^p} dx \right)^{1-s} \left(\int_t^\infty \frac{u(x)}{x} dx \right)^{-s} \left(\int_0^t v^{-\frac{1}{p-1}} \right)^{(p-1)(1-s)}$$

which is bounded by the hypothesis.

The condition $M^{s+p(1-s)}$ is checked in a similar way. □

The next lemma is the crucial step for the extrapolation result. We follow the same ideas as in $\mathcal{A}_p$-theory (see [**GR**], chapter IV]).

LEMMA 2.6. *Let $1 < p < \infty$, $1 \le r < \infty$, $r \ne p$. Let s be such that $\frac{1}{s} = |1 - \frac{r}{p}|$. Let $w \in \mathcal{C}_p$. Then, $\forall u \ge 0$ in $L^s(w)$ there exists $v \ge 0$ in $L^s(w)$ such that:*
a) $u(x) \le v(x)$ for a.e. $x \in (0, +\infty)$.
b) $||v||_{L^s(w)} \le C ||u||_{L^s(w)}$.
c) If $r < p$, $vw \in \mathcal{C}_r$. If $p < r$, $v^{-1} w \in \mathcal{C}_r$.

PROOF. Let us consider first the case $r < p$. Then, if we let $t \in (0,1]$ be such that $r = (1-t)p + t$, we have, $\frac{1}{s} = 1 - \frac{r}{p}$ and $s = p'/t$. We define the sublinear operator $N(u) = \left(S(|u|^{1/t}w)w^{-1}\right)^t$, where S is the Calderón operator. This operator is bounded on $L^{p'/t}(w)$ since $w^{1-p'} = w^{-p'/p} \in \mathcal{C}_{p'}$. We can apply Rubio de Francia's algorithm ([**GR**], lemma 5.1) and given $u \geq 0$ we obtain $v \geq 0$ satisfying a), b) and $N(v) \leq Cv$, that is $S(v^{1/t}w)w^{-1} \leq v^{1/t}$ which means that $v^{1/t}w \in \mathcal{C}_1$. Finally, by lemma 2.5, we have $(v^{1/t}w)^t w^{1-t} \in \mathcal{C}_{t+p(1-t)} = \mathcal{C}_r$.

If $p < r$ the same argument works (duality of the $\mathcal{C}_p$ classes) exactly as in [**GR**], lemma 5.18. □

The previous lemma combined with Hölder's inequality and duality lead us to the following extrapolation result (see [**GR**] theorem 5.19, for details).

PROPOSITION 2.7. *Let T be a sublinear operator acting on functions defined on $(0,+\infty)$. Let $1 \leq r < +\infty$, $1 < p < \infty$. Suppose that T is bounded on $L^r(w)$ (respectively, T is of weak type (r,r)), for every weight $w \in \mathcal{C}_r$ with norm that depends only upon the $\mathcal{C}_r$-constant for w, then T is bounded on $L^p(w)$ (respectively, T is of weak type (p,p)), for all weights $w \in \mathcal{C}_p$ with a norm that depends only upon the $\mathcal{C}_p$-constant for w.*

PROOF. We consider in detail the case where T is assumed to be bounded on $L^r(w)$ for any weight $w \in \mathcal{C}_r$. Let $w \in \mathcal{C}_p$, and for $u \in L^s(w)$, $u \geq 0$, with $\|u\|_{L^s(w)} = 1$, $1/s = 1 - r/p$, let $v \in L^s(w)$ be a function associated to u according to lemma 2.6.

Then, we have the following chain of inequalities

$$\left(\int_0^\infty |Tf(x)|^p w(x)dx\right)^{r/p} = \int_0^\infty |Tf(x)|^r u(x)w(x)dx$$

$$\leq \int_0^\infty |Tf(x)|^r v(x)w(x)dx \leq C\int_0^\infty |f(x)|^r v(x)w(x)dx$$

$$\leq C\left(\int_0^\infty |f(x)|^p w(x)dx\right)^{r/p}\left(\int_0^\infty v(x)^{(p/r)'}w(x)dw\right)^{\frac{1}{(p/r)'}}$$

$$= C\|v\|_{L^s(v)}\|f\|^r_{L^p(w)} \leq C\|f\|^r_{L^p(w)},$$

where we have used duality, Hölder's inverse inequality and the hypothesis. The remaining cases can be obtained in a similar fashion adapting the arguments of [**GR**], Chapter IV, Theorem 5.19, and we shall therefore omit the details. □

REMARK. It is also possible to extrapolate in the case $r = \infty$, using the following extrapolation theorem by García-Cuerva ([**GC**]):

Let S be a positive sublinear operator and let T be a mapping satisfying the following condition: Every time that S is bounded on $L^\infty(v)$, for some $v \geq 0$, T is also bounded on $L^\infty(v)$, with norm depending only on that of S. Let $1 \leq p < \infty$ and $w \geq 0$. Suppose that S is bounded on $L^p(w)$. Then T is also bounded on $L^p(w)$ with norm depending only on that of S.

Therefore, if S is the Calderón operator we have:

PROPOSITION 2.7'. *If T is bounded on $L^\infty(v)$, for any weight $v \in \mathcal{C}_\infty$, then T is bounded on $L^p(w)$, for all $w \in \mathcal{C}_p$ and for all $1 \le p < \infty$.*

It is important for our purposes to compare the classes $\mathcal{C}_p$ with other classes of weights that have appeared in the literature in connection with interpolation theory. According to the factorization theorem we shall restrict ourselves to the case $p = 1$.

i) We recall that a weight is a *quasipower weight* (see [**BK**]) if there exists a positive constant C such that $C^{-1}w \le Sw \le Cw$, (briefly $Sw \sim w$). We say that a weight w is a *functional parameter or a Kalugina weight* if there exist a positive constant C and a $C^{(1}$ positive function φ such that $C^{-1} \le w\varphi \le C$ and $\alpha\varphi(t) \le t\varphi'(t) \le \beta\varphi(t)$, for some $0 < \alpha < \beta < 1$ and for all $t > 0$. An integration by parts shows that a Kalugina weight is a quasipower weight.
The class $\mathcal{C}_1$ is strictly larger than the others. Indeed, let

$$w(t) = \begin{cases} 1/\sqrt{t}, & \text{if } 0 < t \le 1\ ; \\ 1/\sqrt{t-1}, & \text{if } 1 < t \end{cases}$$

It is very easy to compute that

$$Sw(t) = \begin{cases} \pi - 2 + \frac{4}{\sqrt{t}}, & \text{if } 0 < t \le 1; \\ \frac{2}{t} + \frac{2\sqrt{t-1}}{t} + 2\arctan\frac{1}{\sqrt{t-1}}, & \text{if } 1 < t \end{cases}$$

and therefore $w \in \mathcal{C}_1$ but w is not a quasipower weight and consequently it is not a Kalugina weight either (this example is suggested in [**HS**]).

ii) Using the formulas $S = P + Q = P \circ Q = Q \circ P$ it is easy to see that $Sw \sim w$ if and only if $Pw \sim Qw \sim w$. Moreover, in this case, Pw, Qw and even $P^{(k)}w$, $Q^{(k)}w$ are quasipower weights.
If w is a quasipower weight and k is a natural number, then $w(x) = Q^{(k)}w(x)\rho(x)$, where $\rho(x) \sim C$ and $Q^{(k)}w$ is a nonincreasing $\mathcal{C}^{(k-1)}$-function. Therefore we can suppose that w is a nonincreasing very regular function whenever w is a quasipower weight.

iii) Note that $Sf(t) = P(Qf)(t)$, therefore, since Qf is a nonincreasing function (if $f \ge 0$), we see that only the boundedness of P for nonincreasing functions should be considered. Thus, we have that $\mathcal{C}_p = M^p \cap \mathcal{B}_p$ (see Definition 4.1 below).
Furthermore, observe that $M_p \cap M^1 \subseteq \mathcal{C}_p$. The result will follow from the following,

PROPOSITION 2.8. *A weight w satisfies M_p and M^1 if and only if there exists $C > 0$ such that for all $t > 0$*

$$\frac{1}{t}\int_0^t w(x)dx + t^{p-1}\int_t^\infty \frac{w(x)}{x^p}dx \le Cw(t). \tag{2.12}$$

Furthermore any of these two equivalent conditions implies M_p and M^p.

PROOF. Suppose that the weight w satisfies M_p and M^1. Then

$$\begin{aligned}
t^{p-1}\int_t^\infty \frac{w(x)}{x^p}dx &\le Ct^{p-1}\left(\int_0^t w(x)^{-p'/p}dx\right)^{-p/p'} \\
&= C\left(\frac{1}{t}\int_0^t w(x)^{-p'/p}dx\right)^{-p/p'} \\
&\le C\frac{1}{t}\int_0^t w(x)dx \le Cw(t),
\end{aligned}$$

where we have used Jensen's inequality.

In order to conclude the proof we will prove that (2.12) implies the conditions M_p and M^p (note that M^1 is trivially satisfied).

Using arguments similar to the ones used in the proof of the reverse Hölder inequality for the $\mathcal{C}_1$ weights, it is not difficult to prove that condition (2.12) also implies a reverse Hölder inequality. Actually, there exists a constant $C > 0$ and $\epsilon > 0$ such that the following inequality holds

$$\frac{1}{t}\int_0^t x^{-\epsilon}w(x)dx + t^{p-1}\int_t^\infty \frac{x^\epsilon w(x)}{x^p}dx \le Ct^\epsilon w(t).$$

Hence

$$\begin{aligned}
&\left(\int_t^\infty \frac{w(x)}{x^p}dx\right)^{1/p}\left(\int_0^t w(x)^{-p'/p}dx\right)^{1/p'} \\
&\le \left(\int_t^\infty \frac{w(x)}{x^p}dx\right)^{1/p}\left(\int_0^t \left(x^{p-1-\epsilon}\int_x^\infty \frac{y^\epsilon w(y)}{y^p}dy\right)^{-p'/p} dx\right)^{1/p'} \\
&\le C\left(\int_t^\infty \frac{w(x)}{x^p}dx\right)^{1/p}\left(\int_0^t x^{-1+\epsilon p'/p}\left(\int_t^\infty \frac{y^\epsilon w(y)}{y^p}dy\right)^{-p'/p} dx\right)^{1/p'} \\
&\le Ct^{\epsilon/p}\left(\int_t^\infty \frac{w(x)}{x^p}dx\right)^{1/p}\left(\int_t^\infty \frac{y^\epsilon w(y)}{y^p}dy\right)^{-1/p} \le C,
\end{aligned}$$

where in the last inequality we have used the fact that

$$\int_t^\infty \frac{w(x)}{x^p}dx \le t^{-\epsilon}\int_t^\infty \frac{x^\epsilon w(x)}{x^p}dx.$$

Consequently, the condition M_p holds. Now we shall prove that M^p holds:

$$\begin{aligned}
&\left(\int_0^t w(x)dx\right)^{1/p}\left(\int_t^\infty \frac{w(x)^{-p'/p}}{x^{p'}}dx\right)^{1/p'} \\
&\le C\left(\int_0^t w(x)dx\right)^{1/p}\left(\int_t^\infty x^{-1-\epsilon p'/p}\left(\int_0^x y^{-\epsilon}w(y)dy\right)^{-p'/p} dx\right)^{1/p'} \\
&\le Ct^{-\epsilon/p}\left(\int_0^t w(x)dx\right)^{1/p}\left(\int_0^t y^{-\epsilon}w(y)dy\right)^{-1/p} \le C,
\end{aligned}$$

where the estimate

$$\int_0^t y^{-\epsilon}w(y)dy \ge t^{-\epsilon}\int_0^t w(y)dy$$

has been used. □

We conclude this section showing an application of the reverse Hölder type inequality which shows the relation between the classes $\mathcal{C}_p$ for different values of p

PROPOSITION 2.9.
i) If the weight w satisfies the condition M_p then it also satisfies M_q, for all $q > p$ and $\|w\|_{M_q}^q \leq 4^p \|w\|_{M_p}^p$.
ii) If the weight w satisfies the condition M^p then it also satisfies M^q, for all $q > p$ and $\|w\|_{M^q} \leq 8q\|w\|_{M^p}$.
iii) If the weight w satisfies the condition M_p then for all $q > p$, the weight $w(t)t^{q-p}$ satisfies M_q with constant $\|wt^{q-p}\|_{M_p} \leq 8q\|w\|_{M_p}$.

PROOF. Let us remark that the assertion i) is nothing but Hölder's inequality and ii) and iii) are consequences of reverse Hölder type inequalities satisfied by these classes of weights.

i) Let $f \in L^q(w)$. Since $q/p > 1$ we have

$$\int_0^\infty |P(f)|^q w \leq \int_0^\infty P(|f|^{q/p})^p w \leq 4^p \|w\|_{M_p}^p \int_0^\infty |f|^q w.$$

ii) Since $w \in M^p$, the reverse Hölder inequality for this class of weights derived in the Proposition 2.4 implies that for all $\varepsilon > 0$ such that $\epsilon p\|w\|_{M^p} < 1$, the weight $w(t)t^{-\epsilon p} \in M^p$, with constant

$$\|wt^{-\epsilon p}\|_{M^p} \leq \frac{4\|w\|_{M^p}}{1 - \epsilon p\|w\|_{M^p}}.$$

Let $f \in L^q(w)$, with $q > p > 1$, then applying Hölder's inequality with exponents p/q and $(q-p)/q$, we have

$$\begin{aligned}|Q(f)(t)| &\leq \left(\int_t^\infty \frac{|f(x)|^{q/p}}{x^{1-\epsilon q}} dx\right)^{p/q} \left(\int_t^\infty x^{-1-p\epsilon/(q-p)} dx\right)^{(q-p)/q} \\ &\leq C\left[Q(|f(x)|^{q/p} x^{\epsilon q})(t)\right]^{p/q} t^{-\epsilon p},\end{aligned}$$

with $C = \left(\dfrac{q-p}{\epsilon p}\right)^{(q-p)/q}$. Thus,

$$\begin{aligned}\int_0^\infty |Qf|^q w &\leq C^q \int_0^\infty \left[Q(|f|^{q/p} x^\epsilon)(t)\right]^p t^{-\epsilon p} w(t) dt \\ &\leq C' \int_0^\infty \left(|f|^{q/p} x^\epsilon\right)^p x^{-\epsilon p} w(x) dx = C' \int_0^\infty |f(x)|^q w(x) dx,\end{aligned}$$

with

$$C' = \left(\frac{q-p}{\epsilon p}\right)^{q-p} \left(\frac{4\|w\|_{M^p}}{1 - \epsilon p\|w\|_{M^p}}\right)^p.$$

Let $2p\epsilon\|w\|_{M^p} = 1$, then we obtain

$$\|w\|_{M^q} \leq 8\|w\|_{M^p}(q-p)^{1-p/q}.$$

iii) Let f be an element in $L^q(w)$, then $wt^{\epsilon p} \in M_p$ with constant

$$\|wt^{\epsilon p}\|_{M_p} \leq \frac{4\|w\|_{M_p}}{1-\epsilon p\|w\|_{M_p}},$$

whenever $\epsilon p\|w\|_{M_p} < 1$. By Hölder's inequality we have

$$\begin{aligned}
|P(f)(t)| &\leq \frac{1}{t}\int_0^t |f(x)|dx \\
&\leq \left(\frac{1}{t}\int_0^t |f(x)|^{q/p}x^{q/p-1-\epsilon}dx\right)^{p/q}\left(\frac{1}{t}\int_0^t x^{(-1+\epsilon p/(q-p)}dx\right)^{q-p/q} \\
&= C\left(\frac{1}{t}\int_0^t |f(x)|^{q/p}x^{q/p-1-\epsilon}dx\right)^{p/q} t^{-1+p(1+\epsilon)/q},
\end{aligned}$$

with $C = \left(\dfrac{q-p}{\epsilon p}\right)^{(q-p)/q}$. Therefore,

$$\begin{aligned}
\int_0^\infty |P(f)(t)|^q w(t)t^{q-p}dt &\leq C^q \int_0^\infty \left(P\left(|f|^{q/p}x^{q/p-1-\epsilon}\right)(t)\right)^p t^{p\epsilon}w(t)dt \\
&\leq C'' \int_0^\infty |f|^q x^{q-p-\epsilon p}w(x)x^{\epsilon p}dx = C''\int_0^\infty |f|^q x^{q-p}dx,
\end{aligned}$$

where

$$C'' = \left(\frac{q-p}{\epsilon p}\right)^{q-p}\left(\frac{4\|w\|_{M_p}}{1-\epsilon p\|w\|_{M_p}}\right)^p.$$

Thus, if we once again take $2p\epsilon\|w\|_{M_p} = 1$ we obtain

$$\|wt^{q-p}\|_{M_q} \leq 8\|w\|_{M^p}(q-p)^{1-p/q}$$

and the result follows. □

REMARK. In order to complete the information we have about the $\mathcal{C}_p$ weights it is worth to mention here (cf. the Remark after Proposition 3.6 bellow) that the classes $\mathcal{C}_p$ are log-convex. This fact should be added to the results given in the Lemma 2.5, quoted before.

3. Applications to real interpolation: reiteration and extrapolation.

In this section we develop a basic theory of real interpolation associated with Calderón weights. An important new aspect of our theory is the extrapolation theorem 3.6 which plays the role of a reiteration theorem. On the other hand many of the usual properties associated with real interpolation are valid in our context and can be obtained by suitable modifications of known results.

The following properties are easy to prove.

PROPOSITION 3.1. *Let w be a $\mathcal{C}_p$ weight. Then,*

i) The spaces $\bar{A}_{w,p;K}$ and $\bar{A}_{w,p;J}$ are intermediate spaces, i.e., we have $A_0 \cap A_1 \hookrightarrow \bar{A}_{w,p;K} \hookrightarrow A_0 + A_1$ and $A_0 \cap A_1 \hookrightarrow \bar{A}_{w,p;J} \hookrightarrow A_0 + A_1$, $1 \leq p \leq \infty$.

ii) $A_0 \cap A_1$ is always dense in $\bar{A}_{w,p;J}$, $1 \leq p < \infty$.

iii) The spaces $\bar{A}_{w,p;K}$ and $\bar{A}_{w,p;J}$ are interpolation spaces, $1 \leq p \leq \infty$.

iv) $\bar{A}_{w,p;K} \hookrightarrow \bar{A}_{w,p;J}$, $1 \leq p \leq \infty$.

The next important result, due to Brundyi and Krugljak [**BK**], shows that the $\mathcal{C}_p$ conditions are necessary and sufficient for the equivalence between the K and J methods.

PROPOSITION 3.2. *Let $1 \leq p \leq \infty$, then $\bar{A}_{w,p;J} \hookrightarrow \bar{A}_{w,p;K}$ for all compatible pairs of Banach spaces if and only if the weight w is in $\mathcal{C}_p$.*

In [**BK**] the authors showed the necessity of the $\mathcal{C}_p$ condition by means of considering the pair $(L^1(dt), L^1(dt/t))$ plus the technical conditions

$$L^p(w) \hookrightarrow L^1 + L^1(dt/t), \qquad L^p(w) \cap L^{\,\mathrm{loc}}(dt/t) \neq \{0\}$$

(see [**BK**], condition (3.4.2) and Lemma 3.4.4).

However, one could use essentially any pair of rearrangement invariant spaces without any other assumptions. More precisely,

Let $w > 0$ a.e. and let $\bar{A} = (A_0, A_1)$ be a compatible pair of r.i. spaces such that the corresponding fundamental functions satisfy

$$\lim_{s \to 0} \frac{\phi_{A_0}(s)}{\phi_{A_1}(s)} = 0, \qquad \lim_{s \to \infty} \frac{\phi_{A_0}(s)}{\phi_{A_1}(s)} = \infty.$$

(recall that the fundamental function $\phi_{A_0}(s)$ of a r.i. space is given by $\phi_{A_i}(s) = \|\chi_{[0,s]}\|_{A_i}$, $s > 0$, $i = 0, 1$ (see [**BS**])).

If $A_0 \cap A_1 \hookrightarrow \bar{A}_{w,p;J} \hookrightarrow \bar{A}_{w,p;K}$, for some $1 \leq p \leq \infty$ then the weight $w \in \mathcal{C}_p$.

The proof of this fact is left to the reader (see Proposition 4.2. below for a related result).

We consider the duality theory associated with these methods. The following two propositions are straightforward generalizations of the classical results (cf. [**BL**]).

PROPOSITION 3.3. *Let* $w \in \mathcal{C}_p$. *Suppose that* $A_0 \cap A_1$ *is dense in* A_0 *and* A_1. *If* A_i' *denotes the dual space of* A_i, *then* $(A_1', A_0')_{w,p;J} \hookrightarrow (A_0, A_1)'_{w^{-p'/p},p';K}$, $1 < p < \infty$, *and* $(A_1', A_0')_{w^{-1},\infty;J} \hookrightarrow (A_0, A_1)'_{w,1;K}$.

PROPOSITION 3.4. *If the weight* $w \in \mathcal{C}_p$ *then* $(A_0, A_1)'_{w,p;J} \hookrightarrow (A_1', A_0')_{w^{-p'/p},p';K}$, $1 < p < \infty$ *and* $(A_0, A_1)'_{w,1;J} \hookrightarrow (A_1', A_0')_{w^{-1},\infty;K}$

Combining the previous results we get

COROLLARY 3.5. *Suppose that* $w \in \mathcal{C}_p$, *and* $A_0 \cap A_1$ *is dense in* A_0 *and* A_1, *then,*

$$(A_0, A_1)'_{w,p;K} \cong (A_1', A_0')_{w^{-p'/p},p';K} \qquad 1 < p < \infty,$$
$$(A_0, A_1)'_{w,1;K} \cong (A_1', A_0')_{w^{-1},\infty;K}$$

We shall now consider reiteration results. We begin remarking the following fact:

Let $1 \leq p_0, p_1 \leq \infty$ and $w_0 \in \mathcal{C}_{p_0}$, $w_1 \in \mathcal{C}_{p_1}$. Then, the pairs

$$\overline{A}_{\overline{w},\overline{p};K} = (\overline{A}_{w_0,p_0;K}, \overline{A}_{w_1,p_1;K}) \quad \text{and} \quad L_{(\overline{p})}(\overline{w}) = (L^{p_0}(w_0), L^{p_1}(w_1))$$

are almost "bl-pseudoretracts" of each other (in the sense indicated in [**Cw**], p.125]). In order to prove our claim we need to show that the following two statements hold:

1. *Given any* $a \in \Sigma(\overline{A}) = A_0 + A_1$, *let* $K_a(t) = \dfrac{K(a,t;A)}{t}$. *Then, there exists a linear map* V *from* $\Sigma(\overline{A})$ *into the space of measurable functions on* $(0, \infty)$ *such that* $Va = K_a$ *and for all* $b \in \Sigma(\overline{A})$, $|Vb(t)| \leq K_b(t), (t > 0)$ *and, consequently,* $||Vb||_{L_{\overline{p}}(\overline{w})} \leq 2||b||_{\overline{A}_{\overline{p},\overline{w}}}$.

As usual, the last inequality means that V is a bounded operator from $\overline{A}_{w_i,p_i;K}$ into $L^{p_i}(w_i)$, with norm less or equal than 2, $i = 0, 1$.

2. *Given any* $a \in \Sigma(\overline{A}_{\overline{w},\overline{p};K})$, *there exists a linear operator* T *bounded from* $L_{\overline{p}}(\overline{w})$ *into* $\overline{A}_{\overline{w},\overline{p};K}$ *and such that* $T(K_a) = a$.

The proofs of **1** and **2** can be obtained by a straightforward generalization of the methods of [**Cw**] or, alternatively, as a consequence of the general results indicated in [**DO**]. At any rate it is easy to give explicit definitions of the operators S and T, as we now indicate.

If $n \in \mathbb{Z}$, by the Hahn-Banach theorem, there exists a linear functional l_n on $\Sigma(\overline{A})$ such that $l_n(a) = K(2^n, a; \overline{A})$ and $|l_n(b)| \leq K(2^n, b; \overline{A})$, $\forall b \in \Sigma(\overline{A})$. Then, the operator

$$Vb(t) = \sum_{n=-\infty}^{\infty} \frac{l_n(b)}{K(2^n, a; \overline{A})} \cdot \frac{K(t, a; \overline{A})}{t} \chi_{[2^{n-1}, 2^n)}(t)$$

satisfies the required properties.

On the other hand, let $a = \int_0^\infty a(t) \frac{dt}{t}$, with $a(t) \in \Delta(\overline{A}) = A_0 \cap A_1$ and such that $J(t, a(t); \overline{A}) \leq CK(t, a; \overline{A})$, $(t > 0)$ (this is possible thanks to the "fundamental lemma of interpolation"). Then, the operator T given by

$$Tf = \int_0^\infty \frac{f(t)a(t)}{K(t,a;\overline{A})} dt$$

satisfies the required boundedness as it can be easily seen by using the J-functional instead of the K-functional.

The main (and obvious) consequence of **1** and **2** is the following equivalence of K-functionals.

$$K(t,a;\overline{A}_{\overline{w},\overline{p};K}) \sim K(t,K_a;L_{\overline{p}}(\overline{w})), \qquad a \in \Sigma(\overline{A}_{\overline{w},\overline{p};K}). \tag{3.1}$$

Combining **1**, **2** and (3.1) with Sparr's theorem, which establishes that $L_{\overline{p}}(\overline{w})$ is a Calderón pair, lead us to

PROPOSITION 3.6. *$\overline{A}_{\overline{w},\overline{p};K}$ is a Calderón pair. That is, for any $a, b \in \Sigma(\overline{A}_{\overline{w},\overline{p};K})$ with*

$$K(t,b;\overline{A}_{\overline{w},\overline{p};K}) \leq K(t,a;\overline{A}_{\overline{w},\overline{p};K})$$

for all $t > 0$, there exists an operator U bounded in $\overline{A}_{\overline{w},\overline{p};K}$ with $Ua = b$.

Moreover, (3.1) can be also used to prove reiteration results. We consider first a reiteration formula of Holmstedt type. In fact, using the well known characterizations of K-functionals on L^p-spaces (see remark 2 in [**Cw**]), and (3.1) we obtain the following Holmsted type formulae:

For $1 \leq p_0 < p_1 < \infty$, and $\dfrac{1}{\alpha} = \dfrac{1}{p_0} - \dfrac{1}{p_1}$,

$$K(t,a;\overline{A}_{\overline{w},\overline{p},;K}) \sim$$

$$\left(\int_0^{t^\alpha} (K_a(s)(\frac{w_1(s)}{w_0(s)})^{\frac{1}{p_1-p_0}})^{*p_0} ds\right)^{1/p_0} + \left(\int_{t^\alpha}^\infty (K_a(s)(\frac{w_1(s)}{w_0(s)})^{\frac{1}{p_1-p_0}})^{*p_1} ds\right)^{1/p_1}$$

where the nonincreasing rearrangement of the functions is taken with respect to the measure $d\mu = w_0^{\frac{p_1}{p_1-p_0}} w_1^{\frac{-p_0}{p_1-p_0}}$ on $[0,\infty)$.

Although in general this formulae is not easy to decode, (3.1) implies interesting results in particular cases. In fact, an easy consequence of (3.1) is the following reiteration theorem:

$$(\overline{A}_{w_0,p_0;K}, \overline{A}_{w_1,p_1;K})_{w,p;K} = \overline{A}_{(L^{p_0}(w_0),L^{p_1}(w_1))_{w,p;K}} \tag{3.2}$$

where the space on the right hand side consists of the elements $a \in \Sigma(\overline{A})$ for which $K_a \in (L^{p_0}(w_0), L^{p_1}(w_1))_{w,p;K}$. Note that in order for (3.2) to be satisfied it is necessary that the weights w_0 and w_1 are in the $\mathcal{C}_{p_0}$ and $\mathcal{C}_{p_1}$ classes respectively, but it is not relevant whether the weight w belongs or not to the $\mathcal{C}_p$ class.

Consequences of (3.2) are:

i) If we interpolate by the classical real method, then we obtain

$$(\overline{A}_{w_0,p_0;K}, \overline{A}_{w_1,p_1;K})_{\theta,p;K} = \overline{A}_{w,p;K}$$

where $1 \leq p_0, p_1 < \infty$, $w = w_0^{p(1-\theta)/p_0} w_1^{p\theta/p_1}$, $\dfrac{1}{p} = \dfrac{1-\theta}{p_0} + \dfrac{\theta}{p_1}$.

(It suffices to combine (3.2) with known results about the Lions-Peetre method for weighted L_p spaces. See [**BL**]).

REMARK. The Calderón operator is bounded on $L^{p_0}(w_0)$ and on $L^{p_1}(w_1)$. Therefore by interpolation we have that if $w_0 \in \mathcal{C}_{p_0}$ and $w_1 \in \mathcal{C}_{p_1}$ then

$$w = w_0^{p(1-\theta)/p_0} w_1^{p\theta/p_1} \in \mathcal{C}_p$$

where $\frac{1}{p} = \frac{1-\theta}{p_0} + \frac{\theta}{p_1}$.

Making, $p_0 = p_1 = p$ and $0 < \theta < 1$, then we see that $w_0^{1-\theta} w_1^{\theta} \in \mathcal{C}_p$ if $w_0, w_1 \in \mathcal{C}_p$. This shows that, $\mathcal{C}_p$ is a logarithmically convex class (see Lemma 2.5).

ii) (See §**6** below). In the particular case $\overline{A} = (L_1, L_\infty)$, $\overline{A}_{p_0,w_0} = \Lambda(w_0, p_0)$. Then,

$$(\Lambda(w_0, p_0), \Lambda(w_1, p_1))_{\theta,p;K} = \Lambda(w, p)$$

where $w = w_0^{p(1-\theta)/p_0} w_1^{p\theta/p_1}$, $\frac{1}{p} = \frac{1-\theta}{p_0} + \frac{\theta}{p_1}$.

Under our current assumptions it does not matter if we define the Lorentz spaces using f^* or f^{**}. Indeed, since the weights w_0 (resp. w_1) are assumed to be $\mathcal{C}_{p_0}$ (resp. $\mathcal{C}_{p_1}$) weights, it follows that w is a $\mathcal{C}_p$ weight by the previous remark.

iii) As an application of the factorization of $\mathcal{C}_p$ weights we prove the following.

THEOREM 3.7. *Let $1 < p < \infty$ and suppose that $w \in \mathcal{C}_p$. Then, there exist $w_0, w_1 \in \mathcal{C}_1$ such that*

$$(\overline{A}_{w_0,1;K}, \overline{A}_{w_1^{-1},\infty;K})_{1/p',p;K} = \overline{A}_{w,p;K}.$$

PROOF. By (3.2) with $p_0 = 1$ and $p_1 = \infty$,

$$(\overline{A}_{w_0,1;K}, \overline{A}_{w_1^{-1},\infty;K})_{1/p',p;K} = \overline{A}_{(L^1(w_0), L^\infty(w_1^{-1}))_{1/p',p;K}}$$

where $w_0, w_1 \in \mathcal{C}_1$. Then, it is enough to apply the factorization and observe that

$$(L^1(w_0), L^\infty(w_1^{-1}))_{1/p',p;K} = L^p(w_0 w_1^{1-p}). \tag{3.3}$$

This last equality is true since

$$K(t, f; (L^1(w_0), L^\infty(w_1^{-1}))) = K(t, f w_1^{-1}; (L^1(w_0 w_1), L^\infty))$$

and

$$(L^1(w_0 w_1), L^\infty)_{1/p',p;K} = L^p(w_0 w_1).$$

So, $f \in (L^1(w_0), L^\infty(w_1^{-1}))_{1/p',p;K}$ if and only if $f w_1^{-1} \in L^p(w_0 w_1)$ and this means that $f \in L^p(w_0 w_1^{1-p})$ □

REMARK. (3.3) applied to the Calderón operator also shows the implication $w_0, w_1 \in \mathcal{C}_1$ then $w = w_0 w_1^{1-p} \in \mathcal{C}_p$.

Next, we are going to compare our interpolation scales with the ones given by functional parameters.

PROPOSITION 3.8. *Let $w \in \mathcal{C}_1$, and let $\bar{A} = (A_0, A_1)$ be a pair. Then, we have*

1) *Pw, Qw and Sw are equivalent to quasipowers.*

2) $\bar{A}_{w,1;K} = \bar{A}_{Pw,1;K}$.

3) $\bar{A}_{w^{-1},\infty;K} = \bar{A}_{(Pw)^{-1},\infty;K}$.

PROOF.

1) We only prove in detail that Pw is a quasipower, the corresponding proofs for Qw and Sw are similar. Since $w \in \mathcal{C}_1$ we have

$$Pw \leq Cw, \quad Qw \leq Cw.$$

Denote $v = Pw$. Then

$$v = Pw \leq Sw = Q(Pw) = Qv,$$

$$Pv = P^2 w \leq CPw = Cv$$

$$Qv = Q(Pw) = P(Qw) \leq CPw = Cv.$$

Hence $v \sim Qv$ and therefore v is equivalent to a nonincreasing function. Thus,

$$v \leq CPv$$

and we have shown that v is a quasipower.

2) Since $Pw \leq Cw$ we have $\bar{A}_{w,1;K} \subset \bar{A}_{Pw,1;K}$. To prove the reverse containment note that for any nonnegative and nonincreasing function f we have

$$\int fw \leq \int P(f)w = \int fQw \leq \int fPQw$$

but from the proof of (1) we have $PQw = QPw \approx Pw$, therefore we get

$$\int fw \leq C \int fPw.$$

The last inequality applied to $f = \frac{K(t,x;\bar{A})}{t}$ implies the desired result.

3) Since $Pw \leq Cw$ we have $\bar{A}_{(Pw)^{-1},\infty;K} \subset \bar{A}_{w^{-1},\infty;K}$. The converse follows from the fact that for any nonnegative and nonincreasing function f we have

$$\sup(\frac{f}{Pw}) \leq \sup(\frac{P(fww^{-1})}{Pw}) \leq \sup(\frac{f}{w}).$$

□

PROPOSITION 3.9. *Let $1 < p < \infty$ and suppose that $w \in \mathcal{C}_p$. Then there exist quasipower weights v_0, v_1 such that for any pair $\bar{A}$ we have*

$$\bar{A}_{w,p;K} = \bar{A}_{v_0 v_1^{1-p},p;K} = \bar{A}_{Pw,p;K}.$$

PROOF. By Proposition 2.2 there exist $w_0, w_1 \in \mathcal{C}_1$ such that the factorization $w = w_0 w_1^{1-p}$ holds. Therefore by Theorem 3.7 we have $(\bar{A}_{w_0,1;K}, \bar{A}_{w_1^{-1},\infty;K})_{1/p',p;K} = \bar{A}_{w,p;K}$. If we combine the last identification with Proposition 3.8 we get

$$(\bar{A}_{Pw_0,1;K}, \bar{A}_{(Pw_1)^{-1},\infty;K})_{1/p',p;K} = \bar{A}_{w,p;K}. \tag{3.4}$$

Moreover, using once again Proposition 3.8 and Theorem 3.7 we have

$$(\bar{A}_{Pw_0,1;K}, \bar{A}_{(Pw_1)^{-1},\infty;K}))_{1/p',p;K} = \bar{A}_{(Pw_0)(Pw_1)^{1-p},p;K}. \tag{3.5}$$

¿From (3.4) and (3.5) we see that the first identification follows with $v_i = Pw_i, i = 0, 1$, and the last one will follow if we can establish

$$Pw \sim Pw_0(Pw_1)^{1-p}. \tag{3.6}$$

To prove (3.6) we proceed as follows,

$$\begin{aligned} P(w_0 w_1^{1-p})(t) &= \frac{1}{t}\int_0^t \frac{w_0(x)}{w_1(x)^{p-1}}dx \\ &\leq \frac{C}{t}\int_0^t \frac{w_0(x)}{Pw_1(x)^{p-1}}dx \\ &\leq \frac{C}{t}\int_0^t \frac{w_0(x)}{Pw_1(t)^{p-1}}dx \\ &= CPw_0(t)Pw_1(t)^{1-p}. \end{aligned}$$

Conversely, we have

$$\begin{aligned} Pw_0(t)Pw_1(t)^{1-p} &= \left(\frac{1}{t}\int_0^t w_0(x)dx\right)\left(\frac{1}{t}\int_0^t w_1(x)dx\right)^{1-p} \\ &\leq \left(\frac{1}{t}\int_0^t w_0(x)dx\right)\frac{1}{t}\int_0^t w_1(x)^{1-p}dx \\ &= \frac{1}{t}\int_0^t \left(\frac{1}{t}\int_0^t w_0\right) w_1(x)^{1-p}dx \\ &\leq \frac{C}{t}\int_0^t \left(\frac{1}{x}\int_0^x w_0\right) w_1(x)^{1-p}dx \\ &\leq \frac{C}{t}\int_0^t w_0(x)w_1(x)^{1-p}dx = P(w_0 w_1^{1-p})(t). \end{aligned}$$

□

We conclude this section with an extrapolation theorem. To motivate the result let us recall the classical reiteration formula:

$$(\bar{A}_{\theta_0,p_0;K}, \bar{A}_{\theta_1,p_1;K})_{\theta,q;K} = (A_0, A_1)_{\eta,q;K}$$

where $\eta = (1-\theta)\theta_0 + \theta\theta_1$. It follows that the second index is not important for reiteration. Thus, for power weights, that is in the classical Lions-Peetre theory, "extrapolation" in the sense of Rubio de Francia is not interesting. In the setting of $\mathcal{C}_p$ -weights, however, the following extrapolation result holds.

THEOREM 3.10. *Let $\bar{A}$, $\bar{B}$ two compatible pairs of Banach spaces, and let T be a linear operator bounded from $\bar{A}_{w,p;K}$ into $\bar{B}_{w,p;K}$ for some p, $1 \leq p < \infty$, and for all $w \in \mathcal{C}_p$ with norm that depends only upon the $\mathcal{C}_p$-constant for w. Then T is also bounded from $\bar{A}_{v,q;K}$ into $\bar{B}_{v,q;K}$ for any q, $1 < q < \infty$, and for all $v \in \mathcal{C}_q$ with norm that depends only upon the $\mathcal{C}_q$-constant for w.*

PROOF. The idea of the proof is to extrapolate the following inequality. If

$$\int_0^\infty \left(\frac{K(t,Ta;\bar{B})}{t}\right)^p w(t)dt \leq C\int_0^\infty \left(\frac{K(t,a;\bar{A})}{t}\right)^p w(t)dt$$

for any weight $w \in \mathcal{C}_p$ then the same formula is true for any q and any $v \in \mathcal{C}_q$.

We can repeat the argument in the proof of proposition 2.7 applied to the function $\dfrac{K(t,Ta;\bar{B})}{t}$ instead of the function $Tf(x)$ that appears there. □

COROLLARY 3.11. . *Let $\bar{A}$, $\bar{B}$ two compatible pairs of Banach spaces, and let T be a linear operator bounded from $\bar{A}_{w,1;K}$ into $\bar{B}_{w,1;K}$ for all w quasipower weight, with norm that depends only upon the C_1-constant for w. Then T is also bounded from $\bar{A}_{v^q t^{q-1},q;K}$ into $\bar{B}_{v^q t^{q-1},q;K}$ for any q, $1 < q < \infty$, and for all v quasipower weight.*

4. Other classes of weights

The analysis carried out above for the $\mathcal{C}_p$ weights can be extended to other classes of weights. In this section consider in some detail the $\mathcal{B}_p$ classes introduced by Ariño-Muckenhoupt in [**AM**]. Recall that these classes of weights characterize the weights for which the Hardy transform, acting only on decreasing functions, is bounded on $L^p(w)$-spaces.

DEFINITION 4.1. *We shall say that a weight w belongs to the class $\mathcal{B}_p$, $1 \leq p < \infty$, if there exists $C > 0$ such that for all $t > 0$ we have*

$$\int_t^\infty (\frac{t}{x})^p w(x)dx \leq C \int_0^t w(x)dx. \tag{$\mathcal{B}_p$}$$

We say that w belongs to $\mathcal{B}^\infty$, if there exists $C > 0$ such that for all $t > 0$ we have

$$\int_0^t Pw(x)dx \leq C \int_0^t w(x)dx. \tag{$\mathcal{B}^\infty$}$$

We also let $\mathcal{C}B_p = \mathcal{B}_p \cap \mathcal{B}^\infty$.

It follows from [**AM**], [**N2**] and [**Sw**] that

I. *$Pf \in L^p(w)$, $1 \leq p < \infty$, for all $f \in L^p(w)$, with f decreasing, if and only if $w \in \mathcal{B}_p$.*

II. *$Qf \in L^p(w)$, $1 \leq p < \infty$, for all $f \in L^p(w)$ with f decreasing if and only if $w \in \mathcal{B}^\infty$.*

III. *$Sf \in L^p(w)$, $1 \leq p < \infty$, for all $f \in L^p(w)$, with f decreasing, if and only if $w \in \mathcal{C}B_p$.*

To see the import of these weights in interpolation theory let us define $\bar{A}_{w,p;k}$, $1 \leq p \leq \infty$, by

$$\bar{A}_{w,p;k} = \left\{a \in A_0 + A_1; \quad k(t,a;\bar{A}) \in L^p(w(t)dt)\right\},$$

where k is the derivative of the functional K, i.e.

$$K(t,a;\bar{A}) = K(0^+,a;\bar{A}) + \int_0^t k(s,a;\bar{A})ds.$$

We denote by

$$\|a\|_{\bar{A}_{w,p;k}} = \left(\int_0^\infty \left(k(t,a;\bar{A})\right)^p w(t)dt\right)^{1/p},$$

with the corresponding modifications when $p = \infty$. Then, we have

PROPOSITION 4.2. *Suppose $A_0 \cap A_1$ is dense in A_0 and the weight $w \in \mathcal{B}_p$ then $\bar{A}_{w,p;K} = \bar{A}_{w,p;k}$. Moreover let $\bar{A} = (A_0, A_1)$ be a compatible pair of rearrangement invariant spaces such that the corresponding fundamental functions satisfy*

$$\lim_{s\to 0} \frac{\phi_{A_0}(s)}{\phi_{A_1}(s)} = 0, \qquad \lim_{s\to\infty} \frac{\phi_{A_0}(s)}{\phi_{A_1}(s)} = \infty.$$

If for some $1 \le p < \infty$, $\bar{A}_{w,p;k} = \bar{A}_{w,p;K}$ with equivalence of norms, then $w \in \mathcal{B}_p$.

PROOF. Since $A_0 \cap A_1$ is dense in A_0, we have (see for instance [**BS**])

$$\frac{1}{t}K(t,a;\bar{A}) = \frac{1}{t}\int_0^t k(x,a;\bar{A})dx.$$

Furthermore, the function k is non increasing and non negative, therefore the sufficiency part of the result follows from [**AM**].

For the converse, let $a = \chi_{[0,s]} \in A_0 \cap A_1$, $s > 0$. It is easy to see that $K(t,a;\bar{A}) = \min\{\phi_0(s), t\phi_1(s)\}$ and $k(t,a;\bar{A}) = \phi_1(s)\chi_{[0,\frac{\phi_0(s)}{\phi_1(s)}]}$. Then, for some constant $C > 0$

$$C\int_0^{\phi_0(s)/\phi_1(s)} \phi_1(s)^p w(t)dt \ge \int_0^{\phi_0(s)/\phi_1(s)} \phi_1(s)^p w(t)dt + \int_{\phi_0(s)/\phi_1(s)}^\infty \phi_0(s)^p w(t)\frac{dt}{t^p}$$

and thus

$$\left[\frac{\phi_0(s)}{\phi_1(s)}\right]^p \int_{\phi_0(s)/\phi_1(s)}^\infty w(t)\frac{dt}{t^p} \le C\int_0^{\phi_0(s)/\phi_1(s)} w(t)dt.$$

Since the function $\dfrac{\phi_0}{\phi_1}$ is continuous, the condition $\mathcal{B}_p$ holds. □

The k-functional appears naturally in weak type interpolation. This is the subject of our next application. In what follows we assume that $A_0 \cap A_1$ is dense in A_0 and that ρ is a non decreasing positive weight on $(0,\infty)$, such that $\lim_{t\to 0^+} \rho(t) = 0$ and $\lim_{t\to\infty} \rho(t) = \infty$.

PROPOSITION 4.3. *Let $T : A_0 \to \bar{B}_{\rho,\infty;k}$ and $T : A_1 \to B_1$. Suppose that*

i) There exist constants $C, \gamma > 0$ such that for all $b, b' \in B_0 + B_1$

$$k(\gamma t, b + b'; \bar{B}) \le C\left(k(t,b;\bar{B}) + k(t,b';\bar{B})\right)$$

ii) u, v are two weights satisfying: $u \in \mathcal{B}_p$ and $u(\rho(x)) = \dfrac{\gamma v(\gamma x)}{\rho'(x)}$,

Then,

$$T : \bar{A}_{u,p;k} \to \bar{B}_{v,p;k}.$$

PROOF. Given $t > 0$, $a \in A_0 + A_1$ let $a = a_0(t) + a_1(t)$ be a nearly optimal decomposition, i.e.

$$\|a_0(t)\|_{A_0} + \rho(t)\|a_1(t)\|_{A_1} \le 2K(\rho(t), a; \bar{A}).$$

Since k is non increasing we have,

$$\begin{aligned}
k(\gamma t, Ta; \bar{B}) &\le C\left[k(t, Ta_0(t); \bar{B}) + k(t, Ta_1(t); \bar{B})\right] \\
&\le C\left[\frac{1}{\rho(t)}\|Ta_0(t)\|_{\bar{B}_{\infty,\rho;k}} + \frac{1}{t}K(t, Ta_1(t); \bar{B})\right] \\
&\le C\max\{\|T\|_{A_0\to\bar{B}_{\infty,\rho;K}}, \|T\|_{A_1\to B_1}\}\left[\frac{1}{\rho(t)}\|a_0(t)\|_{A_0} + \|a_1(t)\|_{A_1}\right] \\
&\le \frac{C}{\rho(t)}K(\rho(t), a; \bar{A}) \le CP[k(\cdot, a; \bar{A})](\rho(t))
\end{aligned}$$

Taking $L^p(v)$ norms we obtain

$$\begin{aligned}
\left\|k(t, Ta; \bar{B})\right\|_{L^p(v(t)dt)} &\le C\left(\int_0^\infty (P[k(\cdot, a; \bar{A})](\rho(\gamma^{-1}t)))^p v(t)dt\right)^{1/p} \\
&\le C\left(\int_0^\infty P[k(\cdot, a; \bar{A})](s)^p \frac{v(\gamma\rho^{-1}(s))\gamma}{\rho'(\rho^{-1}(s))}ds\right)^{1/p} \\
&= \left(\int_0^\infty P[k(\cdot, a; \bar{A})](s)^p u(s)ds\right)^{1/p} \\
&\le C\left(\int_0^\infty k(s, a; \bar{A})^p u(s)ds\right)^{1/p} \\
&\le C\left\|k(s, a; \bar{A})\right\|_{L^p(u(s)ds)}.
\end{aligned}$$

□

REMARK. Let $T : A_0 \to B_{\rho,\infty;K}$ and $T : A_1 \to B_1$. If we do not assume the validity of assumption (ii) in proposition 4.3, then we get $T : A_{u,p;k} \to \bar{B}_{v,p;K}$.

PROOF. Given $t > 0$, $a \in A_0 + A_1$ let $a = a_0(t) + a_1(t)$ be an optimal decomposition, i.e.

$$\|a_0(t)\|_{A_0} + \rho(t)\|a_1(t)\|_{A_1} \le 2K(\rho(t), a; \bar{A}).$$

Then

$$\begin{aligned}
t^{-1}K(t, Ta; \bar{B}) &\le \frac{1}{t}K(t, Ta_0(t); \bar{B}) + \frac{1}{t}K(t, Ta_1(t); \bar{B}) \\
&\le C\max\{\|T\|_{A_0\to\bar{B}_{\infty,\rho;K}}, \|T\|_{A_1\to B_1}\}\left[\frac{1}{\rho(t)}\|a_0(t)\|_{A_0} + \|a_1(t)\|_{A_1}\right] \\
&\le \frac{C}{\rho(t)}K(\rho(t), a; \bar{A}) \le CP[k(\cdot, a; \bar{A})](\rho(t))
\end{aligned}$$

and the result follows. □

Let us also remark that the classes $\mathcal{C}B_p$ also appear naturally in interpolation theory. To see this let us first recall the following

DEFINITION 4.4. *(cf. [**BS**]) We say that an operator T is of generalized weak types $(1,1)$ and (∞,∞) with respect to Banach pairs $\bar{A}, \bar{B}$, if there exists a constant $C > 0$ such that for all $t > 0$ we have*

$$\frac{1}{t}K(t, Tf; \bar{B}) \le CS(K(\cdot, f; \bar{A}))(t)$$

It follows from the definitions that if T is of generalized weak types $(1,1)$ and (∞,∞) with respect to Banach pairs $\bar{A}, \bar{B}$, then for all $w \in \mathcal{B}_p$ we have

$$T : \bar{A}_{w,p;K} \to \bar{B}_{w,p;K}.$$

Examples 4.5

Now we provide some examples of weights which illustrate how our theory can be applied for building interpolation scales.

The most important examples of weights in the $\mathcal{C}_p$ class are the ones defined by powers which correspond to the Lions-Peetre scales. It is very easy to see that $w(x) = x^\alpha \in \mathcal{C}_p$ if and only if $-1 < \alpha < p-1$. Furthermore $w(x) = x^\alpha(1+\log^+ x)^\beta \in \mathcal{C}_p$ if and only if $-1 < \alpha < p-1$, and $\beta \in \mathbb{R}$.

The reverse Hölder type inequality, the factorization theory of the $\mathcal{C}_p$-weights, and straightforward computations show that if w is a $\mathcal{C}_p$-weight, then $w(x)(1+\log^+ x)^\beta$ and $w(x)(1+\log^+(1/x))^\beta$ are also $\mathcal{C}_p$-weights, for all $\beta \in \mathbb{R}$. Furthermore, if w is a $\mathcal{C}_p$-weight, then there exists $\varepsilon > 0$ such that if ϕ is the function defined by $\phi(x) = x^{-\varepsilon}$, for $x < 1$ and $\phi(x) = x^\varepsilon$, for $x \geq 1$, then $v(x) = w(x)\phi(x)$ is also a $\mathcal{C}_p$-weight.

Not all the $\mathcal{C}_p$-weights are quasi-powers. In the comments after the Proposition (2.7)' we indicate a method for constructing $\mathcal{C}_1$ weights not of quasi-power type. The factorization theorem provides us a method for extending these kind of examples for $p > 1$.

In the same way, one verifies that the classes $\mathcal{C}_p$, $\mathcal{B}_p$ and $\mathcal{C}B_p$ are different. The following assertions are not difficult to prove:

i) $\mathcal{B}_p$ is strictly smaller that $\mathcal{C}B_p$.

More precisely, if $w \in \mathcal{B}_p$ then $w\chi_{[0,a]} \in \mathcal{B}_p \setminus \mathcal{C}B_p$, for all $a > 0$. Also if $-1 < \alpha < p-1$, the weight w defined by

$$w(t) = \begin{cases} x^\alpha, & \text{if } 0 < x \leq 1\ ; \\ x^{-1}, & \text{if } 1 < x \end{cases}$$

belongs to $\mathcal{B}_p \setminus \mathcal{B}_\infty$.

ii) $\mathcal{C}B_p$ is strictly smaller that $\mathcal{C}_p$.

Let $\alpha < 0$ and $\beta > -1$ be real numbers. Consider the weight

$$w(t) = \begin{cases} (x^\alpha - 1)^\beta, & \text{if } 0 < x \leq 1\ ; \\ x^\alpha, & \text{if } 1 < x. \end{cases}$$

Then, if $-1 < \alpha < 0$, and $\beta > p/p'$, we have $w \in \mathcal{C}B_p \setminus \mathcal{C}_p$.

5. Extrapolation of weighted norm inequalities via extrapolation theory

In this section we shall develop a connection with the theory of extrapolation of Jawerth and Milman (cf. [**JM**], [**Mi**]). We shall show that under suitable assumptions it is possible to extrapolate from the classical Lions-Peetre scales to the more general scales considered in this paper.

We shall follow the notation of [**JM**] and [**Mi**] to which we also refer for notation and background. In these works it is shown, among other things, that families of estimates, with rates, can be extrapolated. As a consequence of this theory it is possible to obtain sharp estimates of the K functional of a given operator by analyzing the norm behavior of the operator on intermediate families of spaces. This suggests the project of considering weighted norm estimates with rates.

DEFINITION 5.1. *For a weight w we let*

$$\|w\|_{\mathcal{B}_p} = \inf\{C > 0 : \int_t^\infty \left(\frac{t}{x}\right)^p w(x)dx \le C \int_0^t w(x)dx\}.$$

$$\|w\|_{\mathcal{B}\infty} = \inf\{C > 0 : \int_0^t Pw(x)dx \le C \int_0^t w(x)dx\}.$$

$$\|w\|_{\mathcal{C}B_p} = \|w\|_{\mathcal{B}_p} + \|w\|_{\mathcal{B}\infty}.$$

The dependence on the norms of $\mathcal{A}_p$ weights of weighted $L^p(w)$ norm inequalities for classical operators was recently discussed in [**Bu**]. The dependence of the weighted $L^p(w)$ norm inequalities for the Calderón operator S on the $\mathcal{C}_p$-norms is, of course, well understood as well (see proposition 2.9 above).

Using extrapolation in the sense of [**JM**] we show, under mild extra restrictions, a sharper version of Theorem 3.8.

THEOREM 5.2. *Let $\alpha > 0$, $1 < p < \infty$, and let $\bar{A}$ and $\bar{B}$ be Banach pairs. Suppose that T is a bounded linear operator mapping $T : \bar{A}_{w,p;K} \to \bar{B}_{w,p;K}$, for every $w \in \mathcal{C}_p$, with*

$$\|T\|_{\bar{A}_{w,p;K} \to \bar{B}_{w,p;K}} \le C \|w\|_{\mathcal{C}_p}^\alpha,$$

then for all $1 < q < \infty$, for all $w \in \mathcal{C}B_q$ (in particular for all $w \in \mathcal{C}_q$), we have

$$T : \bar{A}_{w,q;K} \to \bar{B}_{w,q;K}.$$

PROOF. A computation shows that if we let $w_\theta(x) = x^{(1-\theta)p-1}, \theta \in (0,1)$, then

$$\|w_\theta\|_{\mathcal{C}_p} = p^{-1/p} p'^{-1/p'} (\frac{1}{1-\theta} + \frac{1}{\theta}).$$

It follows that

$$T : \bar{A}_{\theta,p;K} \rightarrow \bar{B}_{\theta,p;K}.$$

with

$$\|T\|_{\bar{A}_{\theta,p;K} \rightarrow \bar{B}_{\theta,p:K}} \leq c(1-\theta)^{-\alpha}\theta^{-\alpha}.$$

Therefore by [**JM**] we have

$$K(t, Ta; \bar{B}) \leq c \int_0^t \left(\log \frac{t}{s}\right)^{\alpha-1} K(s, a; \bar{A})\, \frac{ds}{s} + t \int_t^\infty \left(\log \frac{s}{t}\right)^{\alpha-1} \frac{K(s, a; \bar{A})}{s} \frac{ds}{s}$$

and thus,

$$\frac{K(t, Ta; \bar{B})}{t} \leq c \left(P^{(\alpha)} \left(\frac{K(\cdot, a; \bar{A})}{\cdot} \right)(t) + Q^{(\alpha)} \left(\frac{K(\cdot, a; \bar{A})}{\cdot} \right)(t) \right).$$

Now, suppose $\alpha \geq 1$ and let $n \in \mathbb{N}$ such that $n \leq \alpha < n+1$. Then,

$$\int_0^t \left(\log \frac{t}{s}\right)^{\alpha-1} K(s, a; \bar{A})\, \frac{ds}{s}$$

$$= \int_0^{t/e} \left(\log \frac{t}{s}\right)^{\alpha-1} K(s, a; \bar{A})\, \frac{ds}{s} + \int_{t/e}^t \left(\log \frac{t}{s}\right)^{\alpha-1} K(s, a; \bar{A})\, \frac{ds}{s}$$

$$\leq \int_0^{t/e} \left(\log \frac{t}{s}\right)^{n} K(s, a; \bar{A})\, \frac{ds}{s} + \int_{t/e}^t K(s, a; \bar{A})\, \frac{ds}{s},$$

therefore $P^{(\alpha)}$ is controlled by the sum of P and $P^{(n+1)}$.

If $0 < \alpha < 1$, we have

$$\int_0^{t/e} \left(\log \frac{t}{s}\right)^{\alpha-1} K(s, a; \bar{A})\, \frac{ds}{s} \leq \int_0^{t/e} K(s, a; \bar{A})\, \frac{ds}{s},$$

and

$$\int_{t/e}^t \left(\log \frac{t}{s}\right)^{\alpha-1} K(s, a; \bar{A})\, \frac{ds}{s} \leq K(t, a; \bar{A}) \int_{t/e}^t \left(\log \frac{t}{s}\right)^{\alpha-1} \frac{ds}{s}$$

$$= K(t, a; \bar{A}) \int_1^e (\log x)^{\alpha-1} \frac{dx}{x} = CK(t, a; \bar{A}) = C \int_{t/e}^t K(t, a; \bar{A}) \frac{ds}{s}$$

$$\leq C' \int_{t/e}^t \frac{tK(s, a; \bar{A})}{s} \frac{ds}{s} \leq C'' \int_{t/e}^t K(s, a; \bar{A}) \frac{ds}{s}$$

and so $P^{(\alpha)}$ is controlled by P.

Since similar arguments work for Q, and $K(t)/t$ decreases, it follows that for every $1 < q < \infty$, and every $w \in \mathcal{C}B_q$, we have

$$T : \bar{A}_{w,q;K} \rightarrow \bar{B}_{w,q;K}$$

as we wished to show. □

NOTE. The proof of the previous result shows that in practice we can extrapolate from estimates in the Lions-Peetre scales to the more general scales considered in this paper.

REMARKS.
i) We refer to [**Bu**] for a detailed analysis of the dependence on the $\mathcal{A}_p$ norms of w of weighted $L^p(w)$ inequalities for singular integrals, the Hardy-Littlewood maximal operator and other classical operators. These estimates can be also extrapolated in the sense of [**JM**].

ii) There are similar versions of the previous theorem for the other classes of weights.

Example 5.3. We discuss how the methods of this section apply to the study of the maximal operator of Hardy-Littlewood. Let us consider an operator $T : \bar{A}_{\theta,q;K} \to \bar{A}_{\theta,q;K}$ such that

$$\|Tf\|_{\bar{A}_{\theta,q;K}} \leq \frac{c}{\theta} \|f\|_{\bar{A}_{\theta,q;K}}, \theta \in (0,1), q \geq 1.$$

Then the theory of [**JM**] gives

$$K(t, Tf; A_0, A_1) \leq \int_0^t K(s, f; A_0, A_1) \frac{ds}{s}. \tag{5.1}$$

Let us assume, moreover, that the pair $\bar{A}$ is regular and satisfies the following property (an example of pair that satisfies this property is (L^1, L^∞)): there exists a constant $c > 0$, such that for any $x, y \in A_0 + A_1$, we have

$$\frac{d}{dt}K(2t, x+y; A_0, A_1) \leq c\left(\frac{d}{dt}K(t, x; A_0, A_1) + \frac{d}{dt}K(t, y; A_0, A_1)\right), \quad t > 0.$$

We shall also assume that T satisfies the 'weak type estimate'

$$tk(t, Tf; A_0, A_1) \leq c \|f\|_{A_0}$$

where $k(t, ...) = \frac{d}{dt}K(t, ..)$.

Then, we can improve on (5.1) and obtain

$$tk(t, Tf, A_0, A_1) \leq c \int_0^t k(s, f; A_0, A_1) ds \tag{5.2}$$

In fact, given $f \in A_0 + A_1$, let us select an almost optimal decomposition $f = f_0(t) + f_1(t)$ such that

$$K(t, f; A_0, A_1) \sim \|f_0(t)\|_{A_0} + \|f_1(t)\|_{A_1}$$

and moreover such that (cf. [**JM1**]),

$$\|f_0(t)\|_{A_0} \sim K(t, f; A_0, A_1) - tk(t, f; A_0, A_1)$$
$$t\|f_1(t)\|_{A_1} \sim tk(t, f; A_0, A_1).$$

Then,

$$tk(2t, Tf; A_0, A_1) \leq ct\left(k(t, Tf_0(t); A_0, A_1) + k(t, Tf_1(t); A_0, A_1)\right).$$

The weak type estimate gives

$$\begin{aligned} tk(t, Tf_0(t); A_0, A_1) &\leq c\|f_0(t)\|_{A_0} \\ &\leq c\left(K(t, f; A_0, A_1) - tk(t, f; A_0, A_1)\right) \end{aligned}$$

On the other hand the fact that k is decreasing combined with the extrapolation estimate (5.2) gives

$$\begin{aligned} tk(t,Tf_1(t);A_0,A_1) \leq K(t,Tf_1(t);A_0,A_1) &\leq c\int_0^t K(s,f_1(t);A_0,A_1)\frac{ds}{s} \\ &\leq c\int_0^t s\left\|f_1(t)\right\|_{A_1}\frac{ds}{s} \\ &\leq c\int_0^t k(t,f;A_0,A_1)ds \\ &= ctk(t,f;A_0,A_1) \end{aligned}$$

Consequently,

$$\begin{aligned} tk(2t,Tf;A_0,A_1) &\leq c\{K(t,f;A_0,A_1)-tk(t,f;A_0,A_1)\}+c\{tk(t,f;A_0,A_1)\} \\ &\leq cK(t,f;A_0,A_1) = c\int_0^t k(s,f;A_0,A_1)ds \end{aligned}$$

as we wished to show.

In particular, let T be a sublinear operator mapping L^p into L^p for $p \in (1,\infty)$ and such that

$$\left\|Tf\right\|_p \leq \frac{c}{p-1}\left\|f\right\|_p.$$

The previous discussion implies that we have

$$K(t,Tf;L^1,L^\infty) \leq c\int_0^t K(s,f;L^1,L^\infty)\frac{ds}{s}$$

or equivalently, in terms of rearrangements, we have

$$(Tf)^{**}(t) \leq c\int_0^t sf^{**}(s)\frac{ds}{s}$$

To place ourselves under the conditions that apply to the maximal operator of Hardy-Littlewood, let us also assume that T is also an operator of weak type $(1,1)$. Then, we obtain moreover that

$$(Tf)^*(t) \leq cf^{**}(t) = Pf^*(t).$$

At this point we can apply the weighted norm inequalities for the operator P studied in this paper. We also refer to our next example and to Example 6.1 below.

Example 5.4. Let T be an operator such that

$$\left\|Tf\right\|_p \leq c\frac{p^2}{p-1}\left\|f\right\|_p, \ 1<p<\infty \qquad (5.3).$$

It follows from [**JM**] that

$$K(t,Tf;L^1,L^\infty) \leq c\left(\int_0^t K(s,f;L^1,L^\infty)\frac{ds}{s} + t\int_t^\infty \frac{K(s,f;L^1,L^\infty)}{s}\frac{ds}{s}\right),$$

and since $K(t,f,g;L^1,L^\infty) = t\int_0^t g^*(s)ds = tg^{**}(t)$,

we get

$$(Tf)^{**}(t) \leq S(f^{**})(t) \qquad (5.3)'$$

As a consequence we get that for $w \in \mathcal{C}B_p$, $1 < p < \infty$,

$$T : \Lambda(w,p) \to \Lambda(w,p).$$

Thus, we see that the theory of weighted norm inequalities for T acting on weighted Lorentz spaces can be extrapolated from the classical unweighted L^p theory. Moreover, if T is of weak type $(1,1)$ the estimate (5.3)' can be sharpened replacing throughout ** by * leading to

$$(Tf)^*(t) \le cS(f^*)(t).$$

In particular these results apply to the Hilbert transform H, given by

$$Hf(x) = \text{ pv } \int_{-\infty}^{\infty} \frac{f(x-y)}{y} dy.$$

The corresponding rearrangement inequalities here are due to O'Neil-Weiss and Calderón, and the weighted Lorentz norm inequalities follow from the work of Boyd and Ariño-Muckenhoupt. In this sense one should view this example as an extention, via extrapolation, of classical estimates for the Hilbert transform to general operators that satisfy the family of estimates (5.3).

REMARK.

The mechanism discussed in the previous examples can be applied to study weighted Lorentz norm inequalities for other classical operators of analysis (cf. [**S**]). In fact, for many of these operators we have very precise L^p estimates. In particular, this remark applies to weighted versions of Sobolev embedding theorems. We shall leave the particular details to the interested reader. When applied to power-logarithmic weights, as described in Example 4.5, we get an extension of [**BeRu**], where such estimates are extensively studied (cf. also [**EOP**]).

6. Applications to function spaces

In this section we indicate briefly how our results fit in the applications of interpolation theory to the study of specific scales of function spaces. In particular we indicate how our results allow us to extend known results for weighted $L^p(w)$ spaces to other scales of spaces. Some of the material here will be useful in §11 when we consider applications to commutator theorems in the context of concrete scales of function spaces.

Naturally our first example deals with Lorentz spaces.

Example 6.1. For the pair (L^1, L^∞) we have

$$\frac{K(t, f; L^1, L^\infty)}{t} = f^{**}(t)$$
$$k(t, f; L^1, L^\infty) = f^*(t)$$

Therefore,

$$(L^1, L^\infty)_{w,p;k} = \Lambda(w, p)$$

where $\Lambda(w, p) = \{f : \int_0^\infty f^{*p} w < \infty\}$ is the corresponding Lorentz space.

If $w \in \mathcal{B}_p$ we have, according to Proposition 4.2,

$$(L^1, L^\infty)_{w,p;K} = \Lambda(w, p).$$

Using theorem 5.2 we thus obtain an extrapolation theorem for operators acting on Lorentz spaces with $\mathcal{C}B_p$ weights which combines features of the Rubio de Francia and the Jawerth-Milman methods. Likewise, if the rates hypothesized in theorem 5.2 are not available we can use the extrapolation theorem 2.6. Let us also note that, moreover, if additional information is available for the operator under study then we can obtain stronger results. For more on this point see Example 5.3 and 5.4 above and Proposition 5.2.2 of [**JM**].

In connection with the following example we also refer the reader to the reiteration result in §3. *ii*).

Example 6.2. We shall now consider Tent spaces. These spaces, which were introduced by Coifman, Meyer and Stein (cf. [**CMS**]), provide a natural framework to study many problems in harmonic analysis. For the characterization of the interpolation spaces between Tent spaces we refer to [**CMS**], [**AM1**], [**AM2**], and the papers quoted therein. We now show that the weighted versions of these spaces introduced in [**So**] fall out in the scope of our theory.

We start by recalling the definitions. For $x \in \mathbb{R}^n$, let $\Gamma(x) = \{(y,t) \in \mathbb{R}^{n+1}_+ : \|x - y\| < t\}$, and for measurable functions f defined on $\mathbb{R}^{n+1}$, $0 < q < \infty$, let

$$A_q(f)(x) = \left(\int_{\Gamma(x)} |f(y,t)|^q \frac{dydt}{t^{n+1}} \right)^{1/q}$$

and for $q = \infty$ we let

$$A_\infty(f)(x) = \sup_{(y,t)\in\Gamma(x)} |f(y,t)|.$$

For $0 < p < \infty$, $0 < q \le \infty$, the Tent spaces T_q^p are defined by

$$T_q^p = \{f : A_q(f) \in L^p(\mathbb{R}^n)\}.$$

There are several ways to define weighted versions of these spaces. For example, extending somewhat [**So**], we let, for a weight w defined on $\mathbb{R}_+$,

$$T_q\Lambda(p,w) = \{f : A_q(f) \in \Lambda(p,w)\}.$$

Likewise, if w is a weight defined on $\mathbb{R}^n$, we can define weighted versions of Tent spaces associated with $L^p(w)$ spaces, by

$$T_q^p(w) = \{f : A_q(f) \in L^p(w)\}.$$

Before proceeding further let us point out a simple construction with $\mathcal{B}_p$ weights which will be useful to derive reiteration theorems in our setup. Let $1 \le p < r$, then $\nu \in \mathcal{B}_{r/p}$ if and only if there exists $w \in \mathcal{B}_r$ such that $\nu(t) = w(t^{1/p})t^{-1/p'}$. Indeed suppose that $\nu(t) = w(t^{1/p})t^{-1/p'}$ with $w \in \mathcal{B}_r$, then, for $l > 0$, we have

$$\begin{aligned}\int_l^\infty (\frac{l}{t})^{r/p}\nu(t)dt &= p\int_{l^{1/p}}^\infty (\frac{l^{1/p}}{x})^r w(x)dx \le c\int_0^{l^{1/p}} w(x)dx \\ &= c\int_0^l \nu(u)du\end{aligned}$$

and therefore $\nu \in \mathcal{B}_{r/p}$. On the other hand, if $\nu \in \mathcal{B}_{r/p}$, then a similar calculation (cf. [**N2**]) shows that $w(t) = t^{p/p'}\nu(t^p) = t^{p-1}\nu(t^p) \in \mathcal{B}_r$ and, consequently, we can write $\nu(t) = w(t^{1/p})t^{-1/p'}$, with $w \in \mathcal{B}_r$ as we claimed.

We shall now briefly indicate how the classical interpolation theory for Tent spaces can be extended to our setting. Consider the computation of the K functional for certain pairs of Tent spaces. We note that

$$K(t,f;T_q^p, L^\infty(\mathbb{R}^{n+1})) \sim K(t, A_q(f); L^p(\mathbb{R}^n), L^\infty(\mathbb{R}^n))$$

(in fact the argument of [**AM1**] for $q = \infty$, works also for $0 < q \le \infty$). Therefore (cf. [**BL**]), for $1 \le p < r$, $w \in \mathcal{B}_r$, and $w_p(t) = w(t^{1/p})t^{-1/p'}$, we get

$$\left(\frac{K(t,f;T_q^p,L^\infty)}{t}\right)^r \sim \left(\frac{1}{t^p}\int_0^{t^p} A_q(f)^*(s)^p ds\right)^{r/p}$$

and

$$\begin{aligned}\int_0^\infty \left(\frac{K(t,f;T_q^p,L^\infty)}{t}\right)^r w(t)dt &\sim \int_0^\infty P(A_q(f)^{*p})(t^p)^{r/p}w(t)dt \\ &\sim \int_0^\infty P(A_q(f)^{*p})(t)^{r/p}w(t^{1/p})t^{-1/p'}dt \\ &\le c\int_0^\infty A_q(f)^{*r}(t)w_p(t)dt.\end{aligned}$$

Conversely, we observe that

$$\begin{aligned}\int_0^\infty A_q(f)^{*r}(t)w_p(t)dt &= \int_0^\infty \left(A_q(f)^{*p}(t)\right)^{r/p} w_p(t)dt \\ &\leq \int_0^\infty \left(P(A_q(f)^{*p})(t)\right)^{r/p} w_p(t)dt\ . \\ &\sim \int_0^\infty \left(\frac{K(t,f;T_q^p,L^\infty)}{t}\right)^r w(t)dt\end{aligned}$$

We have thus shown that for $w \in \mathcal{B}_r$, $1 \leq p < r$, we have

$$\left(T_q^p, L^\infty\right)_{w,r;K} = T_q\Lambda(r,w_p).$$

The case $p = \infty$ requires some modifications (cf. [**CMS**]).

Example 6.3. In this example we discuss weighted Hardy-Sobolev spaces. A general reference for the material here is [**Lu**]. Let $H^p(\mathbb{R}^n)$ denote the usual Hardy H^p spaces on $\mathbb{R}^n, 0 < p < \infty$. For $\alpha > 0$ we let I^α (the Riesz potential of order $-\alpha$) be defined by $(I^\alpha f)\hat{}(\xi) = c\,|\xi|^\alpha \hat{f}(\xi)$. The Hardy-Sobolev spaces $\dot{W}^\alpha_{H^p}(\mathbb{R}^n)$ are defined by

$$\dot{W}^\alpha_{H^p}(\mathbb{R}^n) = \{f \in S' : I^\alpha f \in H^p(\mathbb{R}^n)\}$$

with

$$\|f\|_{\dot{W}^\alpha_{H^p}(\mathbb{R}^n)} = \|I^\alpha f\|_{H^p(\mathbb{R}^n)}\, .$$

The characterization of the $K-$functional for the pair $(H^p(\mathbb{R}^n), \dot{W}^\alpha_{H^p}(\mathbb{R}^n))$ is well known (cf. [**Lu**]), and we have

$$K(t,f;H^p(\mathbb{R}^n),\dot{W}^k_{H^p}(\mathbb{R}^n)) \approx \omega_k(t^{1/k},f)_{H^p},$$

where $\omega_k(t^{1/k},f)_{H^p}$ denotes the H^p modulus of continuity given by

$$\omega_k(t^{1/k},f)_{H^p} = \sup_{|u|<t^{1/k}} \sum_{j=0}^k (-1)^j \binom{k}{j} \|f(\cdot + ju)\|_{H^p}.$$

Thus the interpolation spaces $(H^p(\mathbb{R}^n), \dot{W}^\alpha_{H^p}(\mathbb{R}^n))_{w,q;K}$ give weighted integrability conditions for these moduli of continuity.

It is interesting to remark here, for future use, that nearly optimal decompositions can be achieved through the use of certain "homogeneous" multipliers. Recall that "homogeneous" multipliers are given by a family of uniformly bounded operators on $H^p(\mathbb{R}^n)$, $\{M_\varepsilon\}_{\varepsilon>0}$, such that $(M_\varepsilon f)\hat{}(\xi) = m(\varepsilon\xi)\hat{f}(\xi)$, where $m \in L^\infty, f \in L^2 \cap H^p$ (cf. [**Lu**]). Under suitable conditions we have that

$$f = \left(f - M_{\varepsilon(t)}f\right) + M_{\varepsilon(t)}f$$

is an optimal decomposition for the computation of $K(t,f;H^p(\mathbb{R}^n),\dot{W}^k_{H^p}(\mathbb{R}^n))$. In particular this applies to the Abel-Poisson homogeneous multipliers given by $(W_t^\alpha f)\hat{}(\xi) = e^{-|t\xi|^\alpha}\hat{f}(\xi)$. Then, in the notation of §7 below, we have

$$D_K(t)f = f - W^\alpha_{t^{1/\alpha}}f$$

$$(I - D_K(t))f = W^\alpha_{t^{1/\alpha}}f.$$

Example 6.4. Hardy spaces with weights can be characterized through the use of weighted Tent spaces. This can be seen, for example, by means of establishing weighted norm inequalities showing the equivalence between the $L^p(w)$ norms of non-tangential maximal functions and square functions. In particular this equivalence holds for weights in the $\mathcal{A}_\infty$ class of Muckenhoupt. For more details we refer to [**Wu**]. In particular this remark allows us to transplant the theory of interpolation of weighted Tent spaces to the setting of weighted Hardy spaces. For a more detailed account of the connection between Hardy spaces and Tent spaces we refer to [**CMS**], see also Example 11.10 below.

Example 6.5. Example 6.1 can be easily generalized to the setting of Sobolev spaces. This follows from the fact that

$$K(t, f; W_1^k(\mathbb{R}^n), W_\infty^k(\mathbb{R}^n)) \sim \sum_{|\alpha|\le k} K(t, D^\alpha f; L^1(\mathbb{R}^n), L^\infty(\mathbb{R}^n)).$$

Therefore if $w \in \mathcal{B}_p$ we obtain

$$(W_1^k(\mathbb{R}^n), W_\infty^k(\mathbb{R}^n))_{w,p;K} = W_{\Lambda(w,p)}^k(\mathbb{R}^n)$$

where

$$W_{\Lambda(w,p)}^k(\mathbb{R}^n) = \{f : D^\alpha f \in \Lambda(w,p)(\mathbb{R}^n), |\alpha| \le k\}$$

provided with its natural norm.

Example 6.6. In a similar fashion we can treat the familiar scales of real interpolation spaces including Besov spaces (cf. [**Bui**]) and non-commutative L^p spaces (cf. [**BL**]). In particular non-commutative weighted Lorentz spaces appear as interpolation spaces between non-commutative L^p spaces. The theory of extrapolation of inequalities developed in this paper thus applies in this set-up. In particular, we believe that our approach would have applications to the estimates of the behavior of singular numbers of integral operators (cf. [**We**]) and the references quoted therein.

Example 6.7. Sometimes it is convenient to replace the K-functional by suitable variants involving powers of the norms. The theory developed in this paper can be readily adapted replacing K-functionals with K_p functionals (cf. [**HP**]). In some cases this remark allows to simplify some calculations in the applications. In particular this applies to the theory of Dirichlet spaces developed by Beurling and Denny (cf. [**BD**]), where the K_2 functional plays an important role. Moreover, as it happens in the theory of Dirichlet spaces, the structure of the optimal decompositions is sometimes easier to relate to the solution of classical variational problems. We shall return to this point in §11 in connection with the theory of commutators. The reader interested in Dirichlet spaces is referred to Example 11.12 below.

7. Commutators defined by the K-method

In this section we shall apply the theory of weighted interpolation spaces to the study of commutators in real interpolation scales. The commutator estimates in the real method of interpolation were originally introduced in [**JRW**]. Since then the subject has been extensively studied and applied in the literature. We refer to [**MR**] for a survey of recent contributions. There are different operators associated with the real methods of interpolation. In this paper we shall focus mainly on the K method but analogous results are valid for the J and E methods. However it is relatively easy, in general, to adapt the methods for the K method to deal with other methods. For a general unified approach to the theory of commutators for the real and complex methods we refer to the forthcoming paper [**CKMR**].

The possibility of proving commutator estimates is due to suitable cancellations that allow one to obtain a more favorable estimate that would otherwise been possible to obtain. Another obstacle to be overcame when dealing with these operations is the fact that they are in general non-linear.

In our development here we shall generalize somewhat the basic operators associated with optimal decompositions introduced in [**JRW**], and show that bounded operators in weighted interpolation scales commute with these operations.

We shall now recall the basic facts and definitions. Let $\bar{A} = (A_0, A_1)$ be a Banach pair, and let $a \in A_0 + A_1$, $t > 0$. We shall say that a decomposition $a = a_0(t) + a_1(t)$ is almost optimal for the K method, if

$$\|a_0(t)\|_0 + t\|a_1(t)\|_1 \leq cK(t, a; \bar{A})$$

where c is a constant whose value is fixed during our discussion, for instance $c = 2$.

It is clear that we can always choose almost optimal decompositions such that that the function from $(0, \infty)$ into $A_0 + A_1$ given by $t \to a_0(t) + a_1(t)$ is measurable. To see this we observe that it is always possible to arrange for these decompositions to be constant over dyadic intervals (cf. §9 below for more details). Moreover, we can also arrange for our almost optimal decompositions to be homogeneous. This means that if we let

$$D_K(t, \bar{A})a = D_K(t)a = a_0(t), \quad (I - D_K(t))a = a_1(t)$$

then for $\lambda \in \mathbb{R}$, we have

$$D_K(t)\lambda a = \lambda a, \quad (I - D_K(t))\lambda a = \lambda a.$$

In what follows we shall always consider decompositions with the properties listed above.

We shall say that a weight w satisfies the integrability condition $(IC)_p$, $1 < p < \infty$, if

$$\int_0^\infty \min\left\{1, \frac{1}{t}\right\}^{p'} w^{-p'/p}(t)dt < \infty. \qquad (IC)_p$$

When this condition is fulfilled, it is possible to define the operator Ω_K for the elements of $\bar{A}_{w,p;K}$, $1 < p < \infty$, as

$$\Omega_{K,\bar{A}}a = \Omega_K a = \int_0^1 D_K(t)a\frac{dt}{t} - \int_1^\infty (I - D_K(t))a\frac{dt}{t} \in A_0 + A_1.$$

Indeed,

$$\begin{aligned}
&\int_0^1 \|D_K(t)a\|_0 \frac{dt}{t} + \int_1^\infty \|I - D_K(t)a\|_1 \frac{dt}{t} \\
&\leq 2\int_0^1 K(t,a;\bar{A})\frac{dt}{t} + 2\int_1^\infty \frac{K(t,a;\bar{A})}{t}\frac{dt}{t} \\
&\leq 2\left(\int_0^1 \left(\frac{K(t,a;\bar{A})}{t}\right)^p w(t)dt\right)^{1/p} \left(\int_0^1 w^{-p'/p}\right)^{1/p'} + \\
&+2\left(\int_1^\infty \left(\frac{K(t,a;\bar{A})}{t}\right)^p w(t)dt\right)^{1/p} \left(\int_1^\infty \frac{w^{-p'/p}}{t^{p'}}dt\right)^{1/p'} < \infty
\end{aligned}$$

Note that for $1 < p < \infty$, $\mathcal{C}_p \subset (IC)_p$, but this condition is not clear for $p = 1$ or even for $\mathcal{C}B_p$ weights. In order to show that also for these kind of weights the element Ωa do exists whenever $a \in \bar{A}_{w,p;K}$, let $1 \leq p < \infty$, and suppose that the weight w is in the class $\mathcal{C}B_p$. Since the function $K(x,a;\bar{A})/x$ is a non increasing function in $L^p(w)$ then

$$\int_0^\infty \left(S\left(\frac{K(x,a;\bar{A})}{x}\right)(t)\right)^p w(t)dt < \infty$$

and so, we have

$$S\left(\frac{K(x,a;\bar{A})}{x}\right)(t) < \infty$$

for all $t > 0$. Thus

$$\begin{aligned}
\infty > S\left(\frac{K(x,a;\bar{A})}{x}\right)(t) &= \frac{1}{t}\int_0^t \frac{K(x,a;\bar{A})}{x}dx + \int_t^\infty \frac{K(x,a;\bar{A})}{x}\frac{dx}{x} \\
&\geq C\left(\frac{1}{t}\int_0^t \|a_0(x)\|_0\frac{dx}{x} + \int_t^\infty \|a_1(x)\|_1\frac{dx}{x}\right),
\end{aligned}$$

in particular for $t = 1$.

Suppose T is a bounded linear operator between the Banach pairs $\bar{A}$ and $\bar{B}$, then given $a \in \bar{A}_{w,p;K}$, we can apply the operators $\Omega_{K,\bar{A}}$ and $\Omega_{K,\bar{B}}$ before and after applying T. This leads to the study of the commutator defined by

$$[T, \Omega_K]a = (T\Omega_{K,\bar{A}} - \Omega_{K,\bar{B}}T)a.$$

It is easy to establish the boundedness of this commutator using, for example, the method of [**JRW**]. We give the details for the sake of completeness.

PROPOSITION 7.1. *Let $1 \leq p < \infty$ and w a weight satisfying some of preceding conditions ($(IC)_p$, for $1 < p < \infty$, or $\mathcal{C}B_1$ for $p = 1$). Then the commutator $[T, \Omega_K]$ is bounded from $\bar{A}_{w,p;K}$ into $\bar{B}_{w,p;J}$.*

PROOF. Let $a = a_0(t) + a_1(t)$, and $Ta = b_0(t) + b_1(t)$ be almost optimal decompositions for a and Ta respectively, i.e.

$$\|a(t)\|_0 + t\|a_1(t)\|_1 \leq 2K(t, a; \bar{A})$$
$$\|b_0(t)\|_0 + t\|b_1(t)\|_1 \leq 2K(t, Ta; \bar{B})$$

Then

$$T\Omega_{K,\bar{A}}a - \Omega_{K,\bar{B}}Ta = \int_0^1 (Ta_0(t) - b_0(t))\frac{dt}{t} - \int_1^\infty (Ta_1(t) - b_1(t))\frac{dt}{t}.$$

Since

$$Ta_0(t) - b_0(t) = -(Ta_1(t) - b_1(t)) = u(t) \in B_0 \cap B_1$$

for all $t > 0$, then

$$T\Omega_{K,\bar{A}}a - \Omega_{K,\bar{B}}Ta = \int_0^\infty u(t)\frac{dt}{t}.$$

Hence

$$\begin{aligned} J(t, u(t); \bar{B}) &\leq \|u(t)\|_0 + t\|u(t)\|_1 \\ &\leq \|Ta_0(t)\|_0 + \|b_0\|_0 + t\|Ta_1(t)\|_1 + t\|b_1\|_1 \\ &\leq 4\|T\|_{\bar{A}\to\bar{B}}K(t, a; \bar{A}), \end{aligned}$$

and so

$$\begin{aligned} \|T\Omega_{K,\bar{A}}a - \Omega_{K,\bar{B}}Ta\|_{\bar{B}_{w,p;J}} &\leq \left(\int_0^\infty \left(\frac{J(t, u(t); \bar{B})}{t}\right)^p w(t)dt\right)^{1/p} \\ &\leq 4\|T\|_{\bar{A}\to\bar{B}}\|a\|_{\bar{A}_{w,p;K}}. \end{aligned}$$

COROLLARY 7.2. *Let $1 \leq p < \infty$, and let $w \in \mathcal{C}_p$. Then, the commutator $[T, \Omega_K]$ is bounded from $\bar{A}_{w,p;K}$ into $\bar{B}_{w,p;K}$.*

REMARK. If $\bar{\Omega}_K$ is another operator defined in $A_0 + A_1$ by using a different almost optimal decomposition and T is the identity operator then, by the Proposition 7.1, the operator $\Omega_K - \bar{\Omega}_K$ is bounded from $\bar{A}_{w,p;K}$ into $\bar{A}_{w,p;J}$. Therefore, if we define the domains of these operators by

$$\text{Dom } \Omega_{K,A} = \{a \in \bar{A}_{w,p;K}; \Omega a \in \bar{A}_{w,p;J}\}$$

with

$$\|a\|_D = \|a\|_{\bar{A}_{w,p;K}} + \|\Omega a\|_{\bar{A}_{w,p;J}},$$

these spaces are independent of the particular operator $\Omega_{K,A}$ selected. Moreover, $\|.\|_D$ is a quasinorm and any bounded linear operator T from $\bar{A}$ into $\bar{B}$ is also bounded from Dom $\Omega_{K,A}$ into Dom $\Omega_{K,B}$ (see [**CJM**]).

We can improve the previous result by using the methods appearing in section 7.2 of [**Mi**]. It will be useful to give a complete proof for future reference in the generalizations we present below.

PROPOSITION 7.3. *Let $w \in \mathcal{C}B_p$, $1 \leq p < \infty$, then the commutator $[T, \Omega_K]$ is bounded from $\bar{A}_{w,p;K}$ into $\bar{B}_{w,p;K}$.*

Moreover

$$\|[T,\Omega]a\|_{p,w,K} \leq \|T\|_{\bar{A}\to\bar{B}}\|w\|_{\mathcal{C}B_p}\|a\|_{p,w,K}.$$

PROOF. We start with the equality (for $t > 0$)

$$\Omega_{K,\bar{A}}a + (\log t)a = \int_0^t D_K(s)a\frac{ds}{s} - \int_t^\infty (I - D_K(s))a\frac{ds}{s} \in A_0 + A_1$$

This gives us for each $t > 0$ a decomposition of $\Omega_{K,\bar{A}}a + (\log t)a$ which can be used to estimate its K-functional by

$$\frac{K(t, \Omega_{K,\bar{A}}a + (\log t)a; \bar{A})}{t} \leq 2S\left(\frac{K(s,a;\bar{A})}{s}\right)(t).$$

Now, we can write

$$\frac{K(t,[T,\Omega]a;\bar{B})}{t}$$

$$\leq \frac{K(t, T(\Omega_{K,\bar{A}}a + (\log t)a); \bar{B})}{t} + \frac{K(t, \Omega_{K,\bar{B}}Ta + (\log t)Ta; \bar{B})}{t}$$

$$\leq 2\|T\|S\left(\frac{K(s,a;\bar{A})}{s}\right)(t) + 2S\left(\frac{K(s,Ta;\bar{B}}{s}\right)(t)$$

and, since $\dfrac{K(s,a;\bar{A})}{s}$ and $\dfrac{K(s,Ta;\bar{B})}{s}$ are decreasing functions, the result follows.

With suitable modifications we can deal with higher order commutators. For $n \in \mathbb{N}$ the operators $\Omega_{K,\bar{A};n}$ associated with the almost optimal decomposition $D_K(t)a$ are defined by

$$\Omega_{K,\bar{A};n}a = \frac{1}{(n-1)!}\left(\int_0^1 (\log t)^{n-1}D_K(t)a\frac{dt}{t} - \int_1^\infty (\log t)^{n-1}(I - D_K(t))a\frac{dt}{t}\right).$$

The corresponding result is given by

PROPOSITION 7.4. *Let $1 \leq p < \infty$. Let w be a weight in the class $\mathcal{C}B_p$. Let $T : \bar{A} \to \bar{B}$ a bounded linear operator. For $n = 0, 1, 2, \dots$ we define*

$$C_n a = \begin{cases} T, & \text{if } n = 0 \\ [T, \Omega_{K;1}]a, & \text{if } n = 1 \\ \dots & \\ [T, \Omega_{K;n}]a + \sum_{k=1}^{n-1} \Omega_{K;k}C_{n-k}a, & \text{if } n \geq 2 \end{cases}$$

Then the operator C_n is bounded from $\bar{A}_{w,p;K}$ into $\bar{B}_{w,p;K}$.

The proof mimics the one given in [**Mi3**] and the weighted norm inequalities for the Calderón operator. We shall skip the proof of Proposition 7.4 here since we shall consider a more general result in Proposition 8.4 in the next section.

8. Generalized commutators

In this section we develop a theory of weighted norm inequalities for commutators associated to the $K-$method of interpolation. Our results will include weighted norm estimates for higher order commutators including those of fractional type.

We need a slight generalization of the theory of weights associated with the Calderón operator which we shall now discuss.

Associated with a non negative, locally integrable function h on $(0,\infty)$ we consider the operators

$$P_h(f)(t) = \int_0^t f(x)h(x)dx = tP(fh)(t), \qquad t > 0$$

$$Q_h(f)(t) = t\int_t^\infty f(x)h(x)\frac{dx}{x} = tQ(fh)(t), \qquad t > 0.$$

The case $h(x) = 1/x$ is obviously the most important one.

Associated with these operators we consider the following classes of weights. Let $1 \leq p < \infty$, then we shall say that a weight w belongs to the class $M_p(h)$ if the operator P_h satisfies

$$\left\|\frac{1}{t}P_h(f)\right\|_{L^p(w)} \leq C\left\|\frac{f}{t}\right\|_{L^p(w)} \tag{8.1}$$

for any function f such that $f(t)/t \in L^p(w)$.

We say that a weight w belongs to the class $M^p(h)$ if the operator Q_h satisfies

$$\left\|\frac{1}{t}Q_h(f)\right\|_{L^p(w)} \leq C\left\|\frac{f}{t}\right\|_{L^p(w)} \tag{8.2}$$

for any function f such that $f(t)/t \in L^p(w)$.

We also define $C_p(h) = M_p(h) \cap M^p(h)$.

These two weighted norm inequalities, which are generalizations of those discussed in §2, are well understood, and can be characterized in terms of conditions similar to M_p and M^p (see, for instance, section 1.3. in [**Ma**]). Indeed, we have: $w \in M_p(h)$, $(1 < p < \infty)$, if and only if there exist a constant $C > 0$ such that for every $t > 0$

$$\left(\int_t^\infty \frac{w(x)}{x^p}dx\right)^{1/p}\left(\int_0^t w(x)^{-p'/p}t^{p'}h(x)^{p'}dx\right)^{1/p'} \leq C,$$

or

$$\int_t^\infty \frac{w(x)}{x}dx \leq C\frac{w(t)}{th(t)},$$

if $p = 1$.

In a similar way we can show that a weight w satisfies the $M^p(h)$ condition, $(1 < p < \infty)$, if and only if there exist a constant $C > 0$ such that for every $t > 0$

$$\left(\int_0^t w(x)dx\right)^{1/p} \left(\int_t^\infty w(x)^{-p'/p} h(x)^{p'} dx\right)^{1/p'} \leq C,$$

or

$$\frac{1}{t}\int_0^t w(x)dx \leq C\frac{w(t)}{th(t)},$$

if $p = 1$.

In particular, if $1 < p < \infty$ and w is a weight $w \in M_p(h)$, then it satisfies the integrability condition

$$\int_0^t w(x)^{-p'/p} t^{p'} h(x)^{p'} dx < +\infty, \quad t > 0, \tag{8.3}$$

and, if $w \in M^p(h)$ then

$$\int_t^\infty w(x)^{-p'/p} h(x)^{p'} dx < +\infty, \quad t > 0. \tag{8.4}$$

We shall say that a weight belongs to the class $IC_p(h)$ if it satisfies conditions (8.3) and (8.4).

In what follows it will be convenient to define the function

$$\nu(t) = \int_1^t h(x)dx$$

which will play here the role that the logarithm plays in the classical theory.

We shall now show that the $\mathcal{C}_p(h)$ also satisfy a kind of reverse Hölder inequality.

PROPOSITION 8.1. *If $w \in \mathcal{C}_p(h)$, then there exists $\epsilon > 0$ such that $we^{\epsilon p \nu} \in M_p(h)$ and $we^{-\epsilon p\nu} \in M^p(h)$.*

PROOF. It is readily verfied by induction that

$$\frac{1}{t}P_h^{(n)}(f)(t) = \frac{1}{t}\int_0^t (\nu(t) - \nu(x))^{n-1} \frac{f(x)h(x)}{(n-1)!}dx$$

and

$$\frac{1}{t}Q_h^{(n)}(f)(t) = \int_t^\infty (\nu(t) - \nu(x))^{n-1} \frac{f(x)h(x)}{(n-1)!}\frac{dx}{x}.$$

Pick a constant $\epsilon > 0$ such $\epsilon C < 1$. It is readily seen that the following operator is well defined,

$$\begin{aligned} R(f)(t) &= \sum_{n=1}^\infty \left(\frac{\epsilon^{n-1}}{t}P_h^{(n)}(f)(t) + \frac{\epsilon^{n-1}}{t}Q_h^{(n)}(f)(t)\right) \\ &= \frac{1}{t}\int_0^t e^{\epsilon(\nu(t)-\nu(x))} f(x)h(x)dx + \int_t^\infty e^{\epsilon(\nu(x)-\nu(t))} f(x)h(x)\frac{dx}{x} \\ &= \int_0^\infty e^{\epsilon|\nu(x)-\nu(t)|} f(x)h(x) \min\left\{\frac{1}{x}, \frac{1}{t}\right\} dx. \end{aligned}$$

Moreover,

$$\|R(f)\|_{L^p(w)} \le \sum_{n=1}^{\infty} \left\| \frac{\epsilon^{n-1}}{t} P_h^{(n}(f) \right\|_{L^p(w)} + \sum_{n=1}^{\infty} \left\| \frac{\epsilon^{n-1}}{t} Q_h^{(n}(f) \right\|_{L^p(w)}$$
$$\le 2 \sum_{n=1}^{\infty} \epsilon^{n-1} C^n \left\| \frac{f}{t} \right\|_{L^p(w)} = C \left\| \frac{f}{t} \right\|_{L^p(w)} .$$

Hence, the operator G defined by $G(f) = R(tf)$ is well defined and bounded, $G : L^p(w) \to L^p(w)$. Writing

$$G(f)(t) = e^{\epsilon\nu(t)} P(xfhe^{-\epsilon\nu(x)})(t) + e^{-\epsilon\nu(t)} Q(xfhe^{\epsilon\nu(x)})(t)$$

we readily obtain the result. □

COROLLARY 8.2. *Let $w \in \mathcal{C}_p(h)$, $1 < p < \infty$ and $\alpha \ge 1$, then*

$$\int_0^t |\nu(x)|^{(\alpha-1)p'} w(x)^{-p'/p} t^{p'} h(x)^{p'} dx < +\infty, \quad t > 0, \tag{8.5}$$

$$\int_t^{\infty} |\nu(x)|^{(\alpha-1)p'} w(x)^{-p'/p} h(x)^{p'} dx < +\infty, \quad t > 0. \tag{8.6}$$

PROOF. Since $we^{\varepsilon p \nu} \in M_p(h)$, we see that condition (8.3) is fullfilled for this new weight. Now it is enough to observe that near to 0,

$$|\nu(t)|^{(\alpha-1)p'} << e^{-\varepsilon p' \nu(t)} = e^{\varepsilon p' |\nu(t)|}.$$

The proof of (8.6) is analogous and uses the second part of proposition 8.1. □

In the framework of commutators of fractional order, we are going to work with $\alpha > 0$. We begin by defining the non linear operator $\Omega_{\nu,\alpha,K,\bar{A}} a = \Omega_{\nu,\alpha}$. Let $a \in A_0 + A_1$, then we let

$$\Omega_{\nu,\alpha} a = \frac{1}{\Gamma(\alpha)} \left[\int_0^1 \nu(t)^{\alpha-1} D_K(t) a h(t) dt - \int_1^{\infty} \nu(t)^{\alpha-1} (I - D_K(t)) a h(t) dt \right]$$

(where $\nu(t)^{\alpha-1} = |\nu(t)|^{\alpha-1} e^{i\ (\alpha-1)\arg\ \nu(t)}$, with $\arg \nu(t) = 0$ or π).

It is easy to check that $\Omega_{\nu,1}$ is well defined on $\bar{A}_{w,p;K}$ whenever the weight $w \in IC_p(h)$ and $1 < p < \infty$. Indeed, we have

$$\int_0^1 \|D_K(t)a\|_0 h(t) dt + \int_1^{\infty} \|(I - D_K(t))a\|_1 h(t) dt$$
$$\le 2 \int_0^1 K(t, a; \bar{A}) h(t) dt + 2 \int_1^{\infty} \frac{K(t, a; \bar{A})}{t} h(t) dt$$
$$\le 2 \left(\int_0^1 \left(\frac{K(t, a; \bar{A})}{t} \right)^p w(t) dt \right)^{1/p} \left(\int_0^1 w(t)^{-p'/p} (th(t))^{p'} \right)^{1/p'} +$$
$$+ 2 \left(\int_1^{\infty} \left(\frac{K(t, a; \bar{A})}{t} \right)^p w(t) dt \right)^{1/p} \left(\int_1^{\infty} w(t)^{-p'/p} h(t)^{p'} dt \right)^{1/p'} < \infty.$$

In a similar way using conditions (8.5) and (8.6) we see that $\Omega_{\nu,\alpha} a$ is well defined for all $\alpha \ge 1$, and for all $a \in \bar{A}_{w,p;J}$, $1 < p < \infty$.

For $p = 1$, we need to use that $w \in \mathcal{C}_1(h)$. If $a \in \bar{A}_{1,w,K}$ the function $K(x,a;\bar{A})/x$ is in $L^1(w)$ and so,

$$\frac{1}{t} P_h^{(n} \left(\frac{K(\cdot, a; \bar{A})}{\cdot} \right)(t) \in L^1(w)$$

for all $n \in \mathbb{N}$. In particular

$$\frac{1}{t} P_h^{(n} \left(\frac{K(\cdot, a; \bar{A})}{\cdot} \right)(t) < \infty$$

for all $n \in \mathbb{N}$ and for all $t > 0$. By induction

$$\frac{1}{t} \int_0^t |\nu(x)|^k \frac{K(x,a;\bar{A})}{x} h(x) dx < \infty$$

far all $t > 0$ and for all $k \in \mathbb{N}$. The same is true for

$$\int_t^\infty |\nu(x)|^k \frac{K(x,a;\bar{A})}{x} h(x) \frac{dx}{x} < \infty.$$

Hence the element $\Omega_{\nu,\alpha} a$ do exists for $\alpha \geq 1$, since the function $|\nu(x)|$ is decreasing on $(0,1)$ and increasing on $(1,\infty)$.

Tha case $0 < \alpha < 1$ is more delicate and the existence of $\Omega_{\nu,\alpha} a$ is a consequence of the Lemma 8.3 (see Proposition 8.4).

Let $1 \leq p < \infty$, $w \in \mathcal{C}_p(h)$ and let $T : \bar{A} \to \bar{B}$ be a bounded linear operator. For $\alpha \geq 0$ we define

$$C_{\nu,\alpha}(T)a = \begin{cases} T, & \text{if } \alpha = 0 \\ [T, \Omega_{\nu,\alpha}]a = (T\Omega_{\nu,\alpha} - \Omega_{\nu,\alpha} T)a, & \text{if } 0 < \alpha \leq 1 \\ [T, \Omega_{\nu,\alpha}]a + \frac{\Omega_{\nu,1}}{\alpha - 1} C_{\nu,\alpha-1} a, & \text{if } 1 < \alpha \leq 2 \\ [T, \Omega_{\nu,\alpha}]a + \frac{2\Omega_{\nu,1}}{\alpha-1} C_{\nu,\alpha-1} a + \frac{2\Omega_{\nu,2}}{(\alpha-1)(\alpha-2)} C_{\nu,\alpha-2} a, & \text{if } 2 < \alpha \leq 3 \\ \dots & \dots \\ [T, \Omega_{\nu,\alpha}]a + \sum_{k=1}^{n-1} \frac{\binom{n-1}{k}}{\binom{\alpha-1}{k}} \Omega_{\nu,k} C_{\nu,\alpha-k} a, & \text{if } n-1 < \alpha \leq n \end{cases}$$

The estimates for the commutators $C_{\nu,\alpha}(T)$ are based on the following

LEMMA 8.3. *Let n be a natural number and let $0 < \theta < 1$, then there exists a constant $C = C(n, \theta, \nu)$ such that for any $a \in \bar{A}_{w,p;K}$ and for all $t > 0$ we have:*

$$\begin{aligned} \int_0^t |\nu(x)|^{\theta-1} |\nu(x) - \nu(t)|^{n-1} K(x,a;\bar{A}) h(x) dx \\ \leq C \int_0^t |\nu(x) - \nu(t)|^{n-1} K(x,a;\bar{A}) h(x) dx \end{aligned} \tag{8.7}$$

$$\begin{aligned} \int_t^\infty |\nu(x)|^{\theta-1} |\nu(x) - \nu(t)|^{n-1} K(x,a;\bar{A}) h(x) \frac{dx}{x} \\ \leq C \int_t^\infty |\nu(x) - \nu(t)|^{n-1} K(x,a;\bar{A}) h(x) \frac{dx}{x}. \end{aligned} \tag{8.8}$$

for all $t > 0$.

PROOF. We give a detailed proof of (8.7); the proof of (8.8) can be obtained by straightforward modifications.
Case 1: $0 < t < 1/2$. It is clear since $|\nu(x)|$ is non increasing in this range.
Case 2: $1/2 \leq t \leq 3/2$.

$$\begin{aligned}\int_{1/3}^{t} &|\nu(x)|^{\theta-1}|\nu(t)-\nu(x)|^{n-1}K(x,a;\bar{A})h(x)dx \\ &\leq |\nu(3/2)-\nu(1/3)|^{n-1}K(3/2,a;\bar{A})\int_{1/3}^{3/2}|\nu(x)|^{\theta-1}h(x)dx \\ &= CK(3/2,a;\bar{A})\end{aligned}$$

Since,

$$\begin{aligned}\int_{1/3}^{t}|\nu(t)-\nu(x)|^{n-1}h(x)dx &\geq \int_{1/3}^{2/5}|\nu(t)-\nu(x)|^{n-1}h(x)dx \\ &\geq |\nu(1/2)-\nu(2/5)|^{n-1}\int_{1/3}^{2/5}h(x)dx = C(>0)\end{aligned}$$

we obtain

$$\begin{aligned}\int_{1/3}^{t} &|\nu(x)|^{\theta-1}|\nu(t)-\nu(x)|^{n-1}K(x,a;\bar{A})h(x)dx \\ &\leq C\int_{1/3}^{t}|\nu(t)-\nu(x)|^{n-1}K(x,a;\bar{A})h(x)dx \\ &\leq C\int_{1/3}^{t}|\nu(t)-\nu(x)|^{n-1}K(x,a;\bar{A})h(x)dx.\end{aligned}$$

On the other hand

$$\begin{aligned}\int_{0}^{1/3} &|\nu(x)|^{\theta-1}|\nu(t)-\nu(x)|^{n-1}K(x,a;\bar{A})h(x)dx \\ &\leq |\nu(1/3)|^{\theta-1}\int_{0}^{1/3}|\nu(t)-\nu(x)|^{n-1}K(x,a;\bar{A})h(x)dx.\end{aligned}$$

Eventually we get the result for this range of t's.
Case 3: $t > 3/2$. We need only observe that

$$\int_{3/2}^{t}|\nu(x)|^{\theta-1}|\nu(t)-\nu(x)|^{n-1}K(x,a;\bar{A})h(x)dx \leq$$

$$\leq \nu(3/2)^{\theta-1}\int_{3/2}^{t}|\nu(t)-\nu(x)|^{n-1}K(x,a;\bar{A})h(x)dx.$$

□

PROPOSITION 8.4. *Let $1 \leq p < \infty$, $0 < \alpha \leq 1$ and $w \in \mathcal{C}_p(h)$. Let $T : \bar{A} \to \bar{B}$ be a bounded linear operator, then the operator $C_{\nu,\alpha} = [T, \Omega_{\nu,\alpha}]$ is bounded from $\bar{A}_{w,p;K}$ into $\bar{B}_{w,p;K}$.*

PROOF. Let $a \in \bar{A}_{w,p;K}$ and $0 < \alpha \leq 1$. A simple computation shows that if $a = a_0(t) + a_1(t)$ is an almost optimal decomposition of the vector a, then

$$\int_0^t |\nu(x)|^{\alpha-1} \|a_0(x)\|_0 h(x)dx \leq 2 \int_0^t |\nu(x)|^{\alpha-1} K(x, a, \bar{A}) h(x)dx$$
$$\leq C \int_0^t K(x, a, \bar{A})(t) h(x)dx < \infty$$

for all $t > 0$.

Since the function $\frac{1}{x}K(x, a; \bar{A})$ is in $L^p(w)$, we have that $\frac{1}{t}P_h(K(\cdot, a; \bar{A}))(t) \in L^p(w)$ and in particular $\frac{1}{t}P_h(K(\cdot, a; \bar{A}))(t) < \infty$, for all $t > 0$.

On the other hand

$$\int_t^\infty |\nu(x)|^{\alpha-1} \|a_1(x)\|_1 h(x)dx < +\infty$$

and so $\Omega_{\nu,\alpha} a \in A_0 + A_1$. Moreover, a simple computation shows that

$$\Omega_{\nu,\alpha} a + \frac{\nu(t)^\alpha}{\Gamma(\alpha+1)} a = \int_0^t \frac{\nu(x)^{\alpha-1}}{\Gamma(\alpha)} a_0(x) h(x)dx - \int_t^\infty \frac{\nu(x)^{\alpha-1}}{\Gamma(\alpha)} a_1(x) h(x)dx.$$

By using the preceding formula we obtain

$$K(t, \Omega_{\nu,\alpha} a + \frac{\nu(t)^\alpha}{\Gamma(\alpha+1)} a; \bar{A}) \leq \int_0^t \frac{|\nu(x)|^{\alpha-1}}{\Gamma(\alpha)} K(t, a_0(x); \bar{A}) h(x)dx$$
$$+ \int_t^\infty \frac{|\nu(x)|^{\alpha-1}}{\Gamma(\alpha)} K(t, a_1(x); \bar{A}) h(x)dx.$$

Then, Lemma 8.3. implies

$$K(t, \Omega_{\nu,\alpha} a + \frac{\nu(t)^\alpha}{\Gamma(\alpha+1)} a; \bar{A}) \leq \frac{C}{\Gamma(\alpha)} \left[P_h(K(x, a; \bar{A}))(t) + Q_h(K(x, a; \bar{A}))(t)\right] \quad (8.9)$$

and therefore

$$\frac{1}{t}K(t, [T, \Omega_{\nu,\alpha}]a; \bar{B}) \leq \|T\|_{\bar{A}\to\bar{B}} \frac{1}{t}K(t, \Omega_{\nu,\alpha,\bar{A}} a + \frac{\nu(t)^\alpha}{\Gamma(\alpha+1)} a; \bar{A})$$
$$+ \frac{1}{t}K(t, \Omega_{\nu,\alpha,\bar{B}} Ta + \frac{\nu(t)^\alpha}{\Gamma(\alpha+1)} Ta; \bar{B}).$$

Hence

$$\|[T, \Omega_{\nu,\alpha}]a\|_{\bar{B}_{w,p;K}} = \left\|\frac{1}{t}K(t, [T, \Omega_{\nu,\alpha}]a; \bar{B})\right\|_{L^p(w)}$$
$$\leq \|T\|_{\bar{A}\to\bar{B}} \left\|\frac{1}{t}K(t, \Omega_{\nu,\alpha,\bar{A}} a + \frac{\nu(t)^\alpha}{\Gamma(\alpha+1)}; \bar{A})\right\|_{L^p(w)}$$
$$+ \left\|\frac{1}{t}K(t, \Omega_{\nu,\alpha,\bar{B}} Ta + \frac{\nu(t)^\alpha}{\Gamma(\alpha+1)} Ta; \bar{B})\right\|_{L^p(w)}$$
$$\leq \frac{\|T\|_{\bar{A}\to\bar{B}}}{\Gamma(\alpha)} \left(\left\|\frac{1}{t}P_h(K(;a; \bar{A}))(t)\right\|_{L^p(w)} + \left\|\frac{1}{t}Q_h(K(;a+; \bar{A}))(t)\right\|_{L^p(w)}\right)$$
$$\leq \frac{1}{\Gamma(\alpha)} \left(\left\|\frac{1}{t}P_h(K(;Ta; \bar{B}))(t)\right\|_{L^p(w)} + \left\|\frac{1}{t}Q_h(K(;Ta+; \bar{B}))(t)\right\|_{L^p(w)}\right)$$

Combining this estimate with (8.1) and (8.2) we finally get

$$\|[T,\Omega_{\nu,\alpha}]a\|_{\bar{B}_{w,p;K}} \le \frac{C\|T\|_{\bar{A}\to\bar{B}}}{\Gamma(\alpha)}\left\|\frac{K(t,a;\bar{A})}{t}\right\|_{L^p(w)}$$
$$= C\|T\|_{\bar{A}\to\bar{B}}\|a\|_{\bar{A}_{w,p;K}}.$$

□

THEOREM 8.5. *Let* $1 \le p < \infty$, $w \in \mathcal{C}_p(h)$ *and let* T *be a bounded linear operator from* $\bar{A}$ *into* $\bar{B}$. *Then, the operator* $C_{\nu,\alpha}$ *is bounded from* $\bar{A}_{w,p;K}$ *into* $\bar{B}_{w,p;K}$, *for any* $0 \le \alpha$.

PROOF.
Let us first consider the case $1 < \alpha \le 2$. It is clear that

$$\Omega_{\nu,\alpha}a + \frac{\nu(t)^\alpha}{\Gamma(\alpha+1)}a = \int_0^t \frac{\nu(x)^{\alpha-1}}{\Gamma(\alpha)}h(x)a_0(x)dx - \int_t^\infty \frac{\nu(x)^{\alpha-1}}{\Gamma(\alpha)}h(x)a_1(x)dx$$
$$\Omega_{\nu,\alpha-1}a + \frac{\nu(t)^{\alpha-1}}{\Gamma(\alpha)}a = \int_0^t \frac{\nu(x)^{\alpha-2}}{\Gamma(\alpha-1)}h(x)a_0(x)dx - \int_t^\infty \frac{\nu(x)^{\alpha-2}}{\Gamma(\alpha-1)}h(x)a_1(x)dx$$
$$\Omega_{\nu,1}a + \nu(t)a = \int_0^t h(x)a_0(x)dx - \int_t^\infty h(x)a_1(x)dx$$

Therefore

$$\Omega_{\nu,\alpha}a - \frac{\nu(t)}{\alpha-1}\Omega_{\nu,\alpha-1}a - \frac{\nu(t)^\alpha}{(\alpha-1)\Gamma(\alpha-1)}a =$$
$$\int_0^t \frac{\nu(x)^{\alpha-2}(\nu(x)-\nu(t))}{\Gamma(\alpha)}h(x)a_0(x)dx - \int_t^\infty \frac{\nu(x)^{\alpha-2}(\nu(x)-\nu(t))}{\Gamma(\alpha)}h(x)a_1(x)dx.$$

Applying the K-functional as before and using Lemma 8.3, we obtain

$$K(t,\Omega_{\nu,\alpha}a - \frac{\nu(t)}{\alpha-1}\Omega_{\nu,\alpha-1}a - \frac{\nu(t)^\alpha}{(\alpha-1)\Gamma(\alpha-1)}a;\bar{A})$$
$$\le C\left(P_h^{(2)}(K(x,a;\bar{A}))(t) + Q_h^{(2)}(K(x,a;\bar{A}))(t)\right).$$

Thus, by denoting $b = [T,\Omega_{\nu,\alpha-1}]a$, we have

$$\frac{1}{t}K(t,C_{\nu,\alpha}a;\bar{B}) \le \frac{1}{t}K(t,T\Omega_{\nu,\alpha}a - \frac{\nu(t)}{\alpha-1}T\Omega_{\nu,\alpha-1}a - \frac{\nu(t)^\alpha}{(\alpha-1)\Gamma(\alpha-1)}Ta;\bar{B})$$
$$+\frac{1}{t}K(t,-\Omega_{\nu,\alpha}Ta + \frac{\nu(t)}{\alpha-1}\Omega_{\nu,\alpha-1}Ta + \frac{\nu(t)^\alpha}{(\alpha-1)\Gamma(\alpha-1)}Ta;\bar{B})$$
$$+\frac{1}{t}K(t,\frac{1}{\alpha-1}\Omega_{\nu,1}b + \frac{\nu(t)}{\alpha-1}\Omega_{\nu,1}b;\bar{B}).$$

The result now follows using the same argument given in the preceding proposition. Indeed, the previous estimate and (8.9) imply that

$$\|C_{\nu,\alpha}a\|_{\bar{B}_{p,w,K}} = \left\|\frac{K(t,C_{\nu.\alpha}a;\bar{B})}{t}\right\|_{L^p(w)}$$
$$\le C\|T\|_{\bar{A}\to\bar{B}}\|\frac{1}{t}P_h^{(2)}(K(x,a;\bar{A}))(t) + \frac{1}{t}Q_h^{(2)}(K(x,a;\bar{A}))(t)\|_{L^p(w)}$$
$$+C\|\frac{1}{t}P_h^{(2)}(K(x,Ta;\bar{B}))(t) + \frac{1}{t}Q_h^{(2)}(K(x,Ta;\bar{B}))(t)\|_{L^p(w)}$$

$$+C\|\frac{1}{t}P_h(K(x,b;\bar{B}))(t)+\frac{1}{t}Q_h(K(x,b;\bar{B}))(t)\|_{L^p(w)}.$$

The first two terms can be controlled by applying (8.1) and (8.2) twice. The remaining term, containing b, can be controlled applying (8.1) and (8.2) as well as Proposition 8.4. The resulting estimate is

$$\|C_{\nu,\alpha}a\|_{\bar{B}_{p,w,K}} \leq C\|a\|_{\bar{A}_{p,w,K}}$$

where the constant C depends only on α, $\|T\|$ and the other constant involved in (8.7) and (8.8).

The remaining cases require slightly more involved computations but the idea remains the same. Let us give complete details for the case $2<\alpha\leq 3$.

It is readily verified that

$$\Omega_{\nu,\alpha}a-\frac{2\nu(t)}{\alpha-1}\Omega_{\nu,\alpha-1}a+\frac{\nu(t)^2}{(\alpha-1)(\alpha-2)}\Omega_{\nu,\alpha-1}a+\frac{2\nu(t)^\alpha}{(\alpha-1)(\alpha-2)\Gamma(\alpha+1)}a$$

$$=\int_0^t\frac{\nu(x)^{\alpha-3}(\nu(x)-\nu(t))^2}{\Gamma(\alpha)}h(x)a_0(x)dx-\int_t^\infty\frac{\nu(x)^{\alpha-3}(\nu(x)-\nu(t))^2}{\Gamma(\alpha)}h(x)a_1(x)dx.$$

The K-functional of this element is now bounded by the corresponding operator $P_h^{(3)}+Q_h^{(3)}$. If we denote by

$$C_{\nu,\alpha-1}a=[T,\Omega_{\nu,\alpha-1}]a+\frac{1}{\alpha-2}\Omega_{\nu,1}[T,\Omega_{\nu,\alpha-2}]a=b$$
$$C_{\nu,\alpha-2}a=[T,\Omega_{\nu,\alpha-2}]a=c$$
$$\rho_\alpha(t)=\frac{2\nu(t)^\alpha}{(\alpha-1)(\alpha-2)\Gamma(\alpha+1)}$$

then

$$K(t,[T,\Omega_{\nu,\alpha}]a+\frac{2}{\alpha-1}\Omega_{\nu,1}b+\frac{2}{(\alpha-1)(\alpha-2)}\Omega_{\nu,2}c;\bar{B})$$

$$\leq K(t,T\Omega_{\nu,\alpha}a-\frac{2\nu(t)}{\alpha-1}T\Omega_{\nu,\alpha-1}a+\frac{\nu(t)^2}{(\alpha-1)(\alpha-2)}T\Omega_{\nu,\alpha-2}a+\rho_\alpha(t)Ta;\bar{B})$$

$$+K(t,-\Omega_{\nu,\alpha}Ta+\frac{2\nu(t)}{\alpha-1}\Omega_{\nu,\alpha-1}Ta-\frac{\nu(t)^2}{(\alpha-1)(\alpha-2)}\Omega_{\nu,\alpha-2}Ta-\rho_\alpha(t)Ta;\bar{B})$$

$$+K(t,\frac{2}{\alpha-1}(\Omega_{\nu,1}b+\nu(t)b)+\frac{2}{(\alpha-1)(\alpha-2)}(\Omega_{\nu,2}c-\nu(t)\Omega_{\nu,1}c-\frac{\nu(t)^2}{\Gamma(3)}c);\bar{B})$$

and we achieve the result as before by following an analogous routine. □

REMARKS.

i) When α is a natural number, we have the usual higher order commutators

$$C_{\nu,n}a=\begin{cases}T, & \text{if } n=0\\ [T,\Omega_{\nu,K;1}]a, & \text{if } n=1\\ \cdots & \\ [T,\Omega_{\nu,K;n}]a+\sum_{k=1}^{n-1}\Omega_{\nu,K;k}C_{\nu,n-k}a, & \text{if } n\geq 2.\end{cases}$$

ii) When $\alpha=1$ and requiring only integrability conditions, we obtain

PROPOSITION 8.6. *Let* $1 \leq p < \infty$. *If the weight* $w \in \mathcal{C}_p(h)$, *then the commutator* $[T, \Omega_{\nu,1}]$ *is bounded from* $\bar{A}_{p,wt^p h^p,K}$ *into* $\bar{B}_{w,p;J}$.

The proof is the same as the one given in proposition 7.1.

9. The quasi Banach case

In this section we indicate the necessary modifications to develop a theory of commutators in the setting of quasi-Banach spaces.

Recall that a quasi-norm on a vector space X is a map $\|\cdot\|: X \to \mathbb{R}^+$ such that

i) $\|x\| > 0 \; \forall \; x \neq 0$.

ii) $\|\lambda x\| = |\lambda| \; \|x\| \; \forall \; \lambda \in \mathbb{R}, x \in X$.

iii) $\exists C \geq 1$ such that $\|x + y\| \leq C(\|x\| + \|y\|) \; \forall \; x, y \in X$.

The definition of an r-norm on X supposes the validity of i), ii) and the replacement of iii) by

iii') $\|x + y\|^r \leq \|x\|^r + \|y\|^r$ for $x, y \in X$ and some $0 < r \leq 1$,

A quasinorm (or an r-norm) defines a metrizable vector topology on X in the usual manner. A vector space X equipped with a quasi-norm (resp. r-norm) is called a quasi-Banach (resp. r-Banach) space if it is complete.

By the concavity of the function t^r, any r-norm is a quasi-norm with $C = 2^{1/r-1}$. Conversely, by the Aoki-Rolewicz theorem (see [**KPR**], [**Ro**]), for any quasi-norm with constant C there exists r, namely $1/r = \log_2(2C)$, and a r-norm $|\cdot|$, such that $|x| \leq \|x\| \leq 4^{1/r}|x|, \; \forall \; x \in X$.

It follows that an r-normed space X is r-Banach if and only if for any given sequence $(x_n)_n$ of vectors in X such that $\sum_{n=1}^{\infty} \|x_n\|^r < \infty$, it follows that $\sum_{n=1}^{\infty} x_n \in X$.

We shall now consider pairs of quasi-Banach spaces (A_0, A_1). Since any r-norm is also an r'-norm for $0 < r' \leq r$, we shall assume from now on that A_0 and A_1 are r-Banach spaces for the same $0 < r \leq 1$.

The definition of the $A_{p,w,K}$ spaces for quasi-Banach spaces is identical to the one given for Banach pairs. However, in order to deal with the J-method and avoid the usual problems related to integration in quasi-Banach spaces, we need to consider discrete versions of the J-method in this context. Since we shall mainly work with the K-method we briefly indicate the necessary changes for this method only. Note that for some constant depending only on r we have

$$K(t, \sum a_n; \bar{A})^r \leq C \sum K(t, a_n; \bar{A})^r.$$

For elements $a \in A_0 + A_1$, we shall consider almost optimal decompositions $a = a_0(t) + a_1(t)$, which are constant over the dyadic intervals $[2^{n-1}, 2^n)$, $n \in \mathbb{Z}$,

$$a_0(t) = \sum_{n \in \mathbb{Z}} a_0(2^{n-1}) \chi_{[2^{n-1}, 2^n)}(t)$$

$$a_1(t) = \sum_{n \in \mathbb{Z}} a_1(2^{n-1}) \chi_{[2^{n-1}, 2^n)}(t)$$

and such that

$$\|a_0(t)\|_0 + t\|a_1(t)\|_1 \le 2K(t, a; \bar{A}),$$

for all $t > 0$.

We only consider here the case $h(x) = 1/x$. It is easy to verify that the operator Ω_K defined by

$$\begin{aligned}\Omega a = \Omega_K a &= \int_0^1 a_0(t)\frac{dt}{t} - \int_1^\infty a_1(t)\frac{dt}{t} \\ &= \left(\sum_{n=-\infty}^{0} a_0(2^{n-1}) - \sum_{n=1}^{\infty} a_1(2^{n-1})\right)\log 2\end{aligned}$$

is well defined on $\bar{A}_{w,p;K}$, for $r < p < \infty$, if w is a weight in the class $\mathcal{C}_p$. In fact, by Hölder's inequality, we have

$$\sum_{n=0}^{\infty} \|a_0(2^{n-1})\|_0^r \log 2 = \int_0^1 \|a_0(t)\|_0^r \frac{dt}{t} \le 2^r \int_0^1 K(t, a; \bar{A})^r \frac{dt}{t}$$

$$\le 2^r \left(\int_0^1 \left(\frac{K(t,a;\bar{A})}{t}\right)^p w(t)dt\right)^{r/p} \left(\int_0^1 w(t)^{-(r/p)(p/r)'} t^{(r-1)(p/r)'} dt\right)^{(p-r)/p}$$

and

$$\sum_{n=1}^{\infty} \|a_1(2^{n-1})\|_1^r \log 2 = \int_1^\infty \|a_1(t)\|_0^r \frac{dt}{t} \le 2^r \int_1^\infty \frac{K(t,a;\bar{A})^r}{t^r}\frac{dt}{t}$$

$$\le 2^r \left(\int_1^\infty \left(\frac{K(t,a;\bar{A})}{t}\right)^p w(t)dt\right)^{r/p} \left(\int_1^\infty w(t)^{-(r/p)(p/r)'} t^{-(p/r)'} dt\right)^{(p-r)/p}$$

where $(p/r)'$ denotes the conjugate index of (p/r). Since by Lemma 2.9 we know that $w(t)t^{p(1/r-1)} \in M_{p/r}$ and $w(t) \in M^{p/r}$, we see that Ω_K is well defined.

More generally in a similar fashion we can show that the operators defined by

$$\begin{aligned}\Omega_\alpha a &= \frac{1}{\Gamma(\alpha)}\left[\int_0^1 (\log t)^{\alpha-1}\frac{a_0(t)}{t}dt - \int_1^\infty (\log t)^{\alpha-1}\frac{a_1(t)}{t}dt\right] \\ &= \frac{(\log 2)^\alpha}{\Gamma(\alpha+1)}\left(\sum_{-\infty}^{0} a_0(2^{n-1})(n^\alpha - (n-1)^\alpha) + \sum_{1}^{\infty} a_1(2^{n-1})(n^\alpha - (n-1)^\alpha)\right)\end{aligned}$$

are well defined for $\alpha > 0$. Indeed, since

$$|n^\alpha - (n-1)^\alpha|^r \le C(\alpha, r)|n^{(\alpha-1)r+1} - (n-1)^{(\alpha-1)r+1}|$$

for all $n \in \mathbb{Z}$, we have that

$$\sum_{n=-\infty}^{0} \|a_0(2^{n-1})\|_0^r |n^\alpha - (n-1)^\alpha|^r \le C\int_0^1 |\log t|^{(\alpha-1)r}\frac{\|a_0(t)\|_0^r}{t}dt$$

$$\le C\|a\|_{p,w,K}^{r/p}\left(\int_0^1 |\log t|^{(\alpha-1)r(p/r)'} w(t)^{-(r/p)(p/r)'} t^{(r-1)(p/r)'} dt\right)^{(p-r)/p}$$

and a similar inequality is true for the other series.

We can now state the following commutator theorem

PROPOSITION 9.1. *Let $0 < r \leq 1$ and $r < p < \infty$. Let w be a weight in the class $\mathcal{C}_p$ and let $\bar{A} = (A_0, A_1)$ be a pair of r-Banach spaces. Then the operator $\Omega_K a = \Omega a$ is well defined in the space $\bar{A}_{w,p;K}$ and, moreover, for any bounded linear operator $T : \bar{A} \to \bar{B}$, the commutator $[T, \Omega]$ is bounded. Furthermore, the operators*

$$C_\alpha a = \begin{cases} T, & \text{if } \alpha = 0 \\ [T,\Omega_\alpha]a = (T\Omega_\alpha - \Omega_\alpha T)a, & \text{if } 0 < \alpha \leq 1 \\ [T,\Omega_\alpha]a + \frac{\Omega_1}{\alpha-1}C_{\alpha-1}a, & \text{if } 1 < \alpha \leq 2 \\ [T,\Omega_\alpha]a + \frac{2\Omega_1}{\alpha-1}C_{\alpha-1}a + \frac{2\Omega_2}{(\alpha-1)(\alpha-2)}C_{\alpha-2}a, & \text{if } 2 < \alpha \leq 3 \\ \dots & \text{if } \dots \\ [T,\Omega_\alpha]a + \sum_{k=1}^{n-1} \frac{\binom{n-1}{k}}{\binom{\alpha-1}{k}}\Omega_k C_{\alpha-k}a, & \text{if } n-1 < \alpha \leq n \end{cases}$$

are bounded from $\bar{A}_{w,p;K}$ into $\bar{B}_{w,p;K}$, where, as before,

$$\Omega_\alpha a = \frac{1}{\Gamma(\alpha)}\left[\int_0^1 (\log)^{\alpha-1}\frac{a_0(t)}{t}dt - \int_1^\infty (\log t)^{\alpha-1}\frac{a_1(t)}{t}dt\right].$$

PROOF. The proof is an adaptation of those of lemma 8.3 and propositions 8.4 and 8.5. Let us sketch the main differences, only in the case $0 < \alpha \leq 1$. The other cases can be proven in a similar fashion.

For any $t \in (0, \infty)$ we have

$$\Omega_\alpha a + \frac{(\log t)^\alpha}{\Gamma(\alpha+1)} = \int_0^t \frac{(\log x)^{(\alpha-1)}}{\Gamma(\alpha+1)}a_0(x)\frac{dx}{x} - \int_t^\infty \frac{(\log x)^{(\alpha-1)}}{\Gamma(\alpha+1)}a_1(x)\frac{dx}{x}.$$

Then, by using a suitable adaptation of lemma 8.3, we obtain

$$\begin{aligned} K(t, \Omega_\alpha a + \frac{(\log t)^\alpha}{\Gamma(\alpha+1)}; \bar{A})^r &\leq C(\alpha,r)\int_0^t |\log x|^{(\alpha-1)r}\|a_0(x)\|^r\frac{dx}{x} \\ &+ C(\alpha,r)t^r\int_t^\infty |\log x|^{(\alpha-1)r}\|a_1(x)\|_1^r\frac{dx}{x} \\ &\leq C(\alpha,r)\int_0^t |\log x|^{(\alpha-1)r}K(x,a;\bar{A})^r\frac{dx}{x} \\ &+ C(\alpha,r)t^r\int_t^\infty |\log x|^{(\alpha-1)r}\frac{K(x,a;\bar{A})^r}{x^r}\frac{dx}{x} \\ &\leq C\int_0^t K(x,a;\bar{A})^r\frac{dx}{x} + Ct^r\int_t^\infty \frac{K(x,a;\bar{A})^r}{x^r}\frac{dx}{x} \\ &\leq CtP\left(\frac{K(x,a;\bar{A})^r}{x}\right)(t) + Ct^rQ\left(\frac{K(x,a;\bar{A})^r}{x^r}\right)(t). \end{aligned}$$

Since

$$\frac{K(x,a;\bar{A})^r}{x} \in L^{p/r}(w(x)x^{p(1/r-1)}), \quad \frac{K(x,a;\bar{A})^r}{x^r} \in L^{p/r}(w)$$

and besides we know that $w(x)x^{p(1/r-1)} \in M_{p/r}$ and $w \in M^{p/r}$ (lemma 2.9) we achieve

$$\left\|t^{-1}K\left(t, \Omega_\alpha a + \frac{(\log t)^\alpha}{\Gamma(\alpha+1)}; \bar{A}\right)\right\|_{L^p(w)}^p$$

$$\leq C \left\| t^{1-r} P\left(\frac{K(x,a;\bar{A})^r}{x}\right)(t)\right\|_{L^{p/r}(w)}^{p/r} + C\left\| Q\left(\frac{K(x,a;\bar{A})^r}{x^r}\right)(t)\right\|_{L^{p/r}(w)}^{p/r}$$

$$\leq C \left\| P\left(\frac{K(x,a;\bar{A})^r}{x}\right)(t)\right\|_{L^{p/r}(wt^{p(1/r-1)})}^{p/r} + C\left\| Q\left(\frac{K(x,a;\bar{A})^r}{x^r}\right)(t)\right\|_{L^{p/r}(w)}^{p/r}$$

$$\leq C \left\| \frac{K(x,a;\bar{A})^r}{x}\right\|_{L^{p/r}(wx^{p(1/r-1)})}^{p/r} + C\left\| \frac{K(x,a;\bar{A})^r}{x^r}\right\|_{L^{p/r}(w)}^{p/r}$$

$$= C \left\| \frac{K(x,a;\bar{A})}{x}\right\|_{L^{p}(w)}^{p} = C\|a\|_{p,w,K}^p.$$

Now the proof follows as that in proposition 8.4. □

We refer the reader to §10 for examples and applications of these results to harmonic analysis.

10. Applications to Harmonic Analysis

In this section we consider applications of our results in harmonic analysis. In particular we consider commutator theorems with multiplication operators in different settings. We also relate the construction of these commutators to the solution of classical variational problems. Our first example deals with the connection of $\mathcal{A}_p$ weights, the space BMO and commutator estimates.

Example 10.1.

The classical example connecting $\mathcal{A}_p$ weights in $\mathbb{R}^n$, BMO functions, and interpolation is given by the commutators associated with the pair $(L^p(w_0), L^p(w_1))$ (see [**JRW**] or example 64 in [**Mi**]). An $\Omega_{K,n}$ associated with this pair is

$$\Omega_{K,n} f(x) = -\frac{1}{n! p^n} f(x) \left(\log \frac{w_0(x)}{w_1(x)} \right)^n, \quad n = 1, 2, 3, \ldots$$

Moreover, if we consider operators of fractional order we have

$$\Omega_{K,\alpha} f(x) = -\frac{1}{\Gamma(\alpha+1) p^\alpha} f(x) \left(\log \frac{w_0(x)}{w_1(x)} \right)^\alpha, \quad \alpha > 0.$$

Let $1 < p < \infty$. If we take $b \in \mathrm{BMO}(\mathbb{R}^n)$ with BMO norm small enough, then $w_0 = e^{-bp/2}$, $w_1 = e^{bp/2}$ are $\mathcal{A}_p$ weights. Therefore, if T is a Calderón-Zygmund operator then T is bounded on $L^p(w_0)$ and on $L^p(w_1)$. Hence $[T, M_b]$ (where $M_b f = bf$) is bounded on $(L^p(w_0), L^p(w_1))_{q,w,K}$ for $1 \leq q < \infty$ and $w \in \mathcal{C}B_q$ (cf. Proposition 3.3).

Combining these considerations with

$$(L^p(w_0), L^p(w_1))_{1/2,p;K} = L^p.$$

we get the following

COROLLARY 10.2. *Let $p > 1$. Let $b \in BMO(\mathbb{R}^n)$ and T any Calderón-Zygmund operator. Then the commutator $[T, M_b]$ is bounded in $L^p(\mathbb{R}^n, v)$ for the following weights in $\mathbb{R}^n$*

$$v(x) = e^{pb(x)/2} \int_0^{e^{-b(x)}} w(t)dt + e^{-pb(x)/2} \int_{e^{-b(x)}}^\infty w(t) t^{-p} dt$$

whenever the weight w satisfies the condition $\mathcal{C}B_p$.

PROOF. It suffices to note that $(L^p(e^{-pb/2}), L^p(e^{pb/2}))_{p,w;K} = L^p(v)$. □

In the fractional case, if we take $1 < \alpha \leq 2$, for example, we have

$$C_\alpha f = \frac{1}{\Gamma(\alpha+1)} T(fb^\alpha) - \frac{1}{\Gamma(\alpha)} bT(fb^{\alpha-1}) + (\frac{1}{\Gamma(\alpha)} - \frac{1}{\Gamma(\alpha+1)}) b^\alpha T(f).$$

Thus, we obtain

COROLLARY 10.3. *Let $p > 1$ and $1 < \alpha \leq 2$. Let $b \in BMO(\mathbb{R}^n)$ and let T be a Calderón-Zygmund operator with kernel $K(x,y)$. Then, the operator T_α with kernel*

$$K_\alpha(x,y) = K(x,y)\left(\frac{1}{\Gamma(\alpha+1)}(b^\alpha(y) - b^\alpha(x)) + \frac{1}{\Gamma(\alpha)} b(x)(b^{\alpha-1}(x) - b^{\alpha-1}(y))\right)$$

is bounded on L^p.

Let h be a locally integrable function on $(0,\infty)$ with $h(t) > 0$ a.e. and $\nu(t) = \int_1^t h(x)dx$. If $t^{-1+p/2}$ satisfies the condition $\mathcal{C}_p(h)$ we have that $[T, M_g]$ is bounded on L^p where $g(x) = \nu(w_0(x)^{1/p} w_1(x)^{-1/p})$, with $w_0, w_1 \in \mathcal{A}_p$.

More precisely, if $h(t) \leq C/t$ for all $t > 0$ then it is easy to see that $\mathcal{C}_p \subset \mathcal{C}_p(h)$. For instance, the function $h(t) = \dfrac{1}{t(1 + |\log t|)}$ satisfies the required condition and so we have

COROLLARY 10.4. *Let $1 < p < \infty$ and let $b \in BMO$ a function with norm small enough and let T be a Calderón-Zygmund operator. Then, $[T, M_g]$ is bounded on L^p for $g(x) = \log(1 + |b(x)|) sgn\, b(x)$.*

Example 10.5.

We shall now develop weighted norm estimates for Jacobians and other operations with suitable cancellations. Our setting will be the one of [**CLMS**]. We also refer the reader to [**Ch**] for background information and a treatment of the unweighted case using paracommutators.

Let S and T be classical singular integral operators acting on functions defined on $\mathbb{R}^n$ and satisfying $T^* \circ S = S^* \circ T$. Let $p \in (1,\infty)$, and let $b \in$ BMO, then for $f \in L^p$ and $g \in L^{p'}$ we have

$$|\langle b, Tf.Sg - Sf.Tg\rangle| \leq C\|f\|_p \|g\|_{p'} \|b\|_{BMO}.$$

Indeed, we formally have

$$\langle b, Tf.Sg - Sf.Tg\rangle = \langle f, T^*(Sg.M_b) - S^*(Tg.M_b)\rangle$$

Now, we write

$$T^*(Sg.M_b) = -T^*([S, M_b])(g) + T^*(S(gM_b))$$

and

$$-S^*(Tg.M_b) = S^*([T, M_b])(g) - S^*(T(gM_b)$$

and therefore we obtain

$$\begin{aligned}\langle b, Tf.Sg - Sf.Tg\rangle = &-\langle f, T^*([S, M_b])(g)\rangle + \langle f, S^*([T, M_b])(g)\rangle \\ &+ \langle f, (T^* \circ S)(gM_b) - (S^* \circ T)(gM_b)\rangle.\end{aligned}$$

The first two terms are under control, by the commutator theorem discussed in the previous example, while the third term is zero by our assumptions on the operators T and S. In view of the duality between H^1 and BMO the result can be also rephrased as

$$\|Tf.Sg - Sf.Tg\|_{H^1} \leq C\|f\|_p\|g\|_{p'}.$$

In particular the previous considerations apply to Jacobians. Indeed, let $f : \mathbb{R}^n \to \mathbb{R}^n$ be a smooth map i.e. $Df \in L^n$ then a result of [**CLMS**] states that the Jacobian $\det(Df) \in H^1$. Let us show this in the case $n = 2$ we write $f = (f_1, f_2)$,

$$Df = \begin{vmatrix} \frac{\partial f_1}{\partial x_1} & \frac{\partial f_1}{\partial x_2} \\ \frac{\partial f_2}{\partial x_1} & \frac{\partial f_2}{\partial x_2} \end{vmatrix},$$

let $(R_j g)\hat{}(\xi) = -i\frac{\xi_j}{|\xi|}\hat{g}(\xi)$, $j = 1, 2$, denote the Riesz transforms, and let $(Gh)\hat{}(\psi) = |\psi|\hat{h}(\psi)$. Then, $\dfrac{\partial f_i}{\partial x_j} = R_j G f_i, i, j = 1, 2$, with $Gf_i \in L^2$. Therefore, since the Riesz transforms are antisymmetric and commute with each other, the previous discussion applied to (R_1, R_2) shows that

$$J(f) = \frac{\partial f_1}{\partial x_1}\frac{\partial f_2}{\partial x_2} - \frac{\partial f_1}{\partial x_2}\frac{\partial f_2}{\partial x_1} \in H^1(\mathbb{R}^2).$$

We should also note that in our setup we can replace singular integral operators by any pair of operators S, T, which act on $L^p(w)$ for all $w \in \mathcal{A}_p$, $1 < p < \infty$, and such that the cancellation condition $T^* \circ S = S^* \circ T$ is satisfied. Using a variant of corollary 10.2 we can derive weighted Lorentz norm inequalities for these operations. Furthermore, using the properties of $\mathcal{A}_p$ weights, the theory of commutators will also produce weighted $\mathcal{A}_p$ norm inequalities for these operations (cf. [**Mi3**]).

Example 10.6.

We consider weighted norm inequalities for commutators with $\Omega_K f = f \log r_f$ or $\Omega_E f = f \log |f|$, which complement recent work by Pérez [**Pe**]. For simplicity, suppose that T is a bounded operator acting on the pair (L^1, L^∞), then (cf. [**JRW**]) the commutator $[T, \Omega_K]$ is bounded on L^p, $1 < p < \infty$. The method of proof of proposition 7.3 and the fact that we have a concrete description of the K-functional for the pair (L^1, L^∞) leads to the estimate

$$[T, \Omega_K]f^{**}(t) \leq c\left(\frac{1}{t}\int_0^t f^{**}(s)\frac{ds}{s} + \int_t^\infty f^{**}(s)\frac{ds}{s}\right).$$

In particular it follows that

$$\|[T, \Omega_K]f\|_{\Lambda(w,p)} \leq c\,\|f\|_{\Lambda(w,p)}$$

for all $w \in \mathcal{B}_p, 1 < p < \infty$, where $r_f(x)$ is the rank function defined in example 10.7 below. Similarly we can obtain weighted norm inequalities for higher order commutators using the rearrangement inequalities implicit in the proof of proposition 8.4. Using the notation of this proposition it follows that

$$\|C_{\nu,\alpha} f\|_{\Lambda(w,p)} \leq c\,\|f\|_{\Lambda(w,p)}.$$

In order to obtain weighted Lorentz norm inequalities for commutators with the operator $\Omega_E f = f \log |f|$ we require the use the E-method of interpolation. Since we have not considered this method in this paper we just indicate the necessary steps (with references) and leave the details for the interested reader. The corresponding E-functional inequalities for commutators with bounded operators T on the pair (L^1, L^∞) (cf. [**Mi**]) give

$$E(t, [T, \Omega_E]f, L^\infty, L^1) \le c \int_t^\infty E(s, f, L^\infty, L^1) \frac{ds}{s}.$$

To interpret this inequality recall that $E(t, h, L^\infty, L^1) = \int_t^\infty \lambda_f(s) ds$ (cf. [**JM1**]), therefore the estimate takes the form

$$\int_t^\infty \lambda_{[T,\Omega_E]f}(s) ds \le c \int_t^\infty \int_s^\infty \lambda_f(u) du \frac{ds}{s}.$$

Weighted Lorentz norm inequalities follow from this estimate using the (equivalent) distribution function description of the norm of Lorentz spaces.

Example 10.7.

We consider the commutators associated to the pair $(L^{1,\infty}, L^\infty)$. For this pair, the K-functional is given by the expression

$$K(t, f; L^{1,\infty}, L^\infty) = \sup_{0<s<t} s f^*(s)$$

where f^* denotes the non increasing rearrangement of f (cf. [**Kr**], [**BL**]). We can consider two different almost optimal decompositions in order to compute the corresponding operators Ω's. The first one is given by $f = g_1 + h_1$, $g_1 = f\chi_{A_t} \in L^{1,\infty}$ and $h_1 = f - g_1 \in L^\infty$, where $A_t = \{x; |f(x)| > f^*(t)\}$. In this case

$$\Omega_\alpha^{(1)} f(x) = \frac{-1}{\Gamma(\alpha+1)} f(x) (\log r_f(x))^\alpha$$

where r_f is the rank function defined by

$$r_f(x) = |\{y; |f(y)| \ge |f(x)|\}|$$

(see, [**CJMR**], [**Mi**]). Indeed

$$\Omega_\alpha^{(1)} f(x) = \frac{1}{\Gamma(\alpha)} \int_0^1 (\log t)^{\alpha-1} f(x) \chi_{A_t}(x) \frac{dt}{t} - \int_1^\infty (\log t)^{\alpha-1} f(x) \chi_{A_t^c}(x) \frac{dt}{t}.$$

Since $t < r_f(x)$ implies that $f^*(t) \ge |f(x)|$ and also $t > r_f(x)$ implies that $f^*(t) < |f(x)|$ we have that

$$\Omega_\alpha^{(1)} f(x) = \frac{1}{\Gamma(\alpha)} \int_{r_f(x)}^1 (\log t)^{\alpha-1} f(x) \frac{dt}{t}$$

for $r_f(x) < 1$ and

$$\Omega_\alpha^{(1)} f(x) = \frac{-1}{\Gamma(\alpha)} \int_1^{r_f(x)} (\log t)^{\alpha-1} f(x) \frac{dt}{t}$$

for $r_f(x) > 1$, so clearly the formula holds.

We can also consider optimal decompositions given by $f = g_2 + h_2$, where $g_2 = (f - f^*(t))\chi_{A_t} \in L^{1,\infty}$ and $h_2 = f - g_2 \in L^\infty$. It is quite easy to see that the corresponding operator associated with this decomposition, $\Omega^{(2)}$, is given by

$$\Omega_{\alpha}^{(2)} f(x) = \Omega_{\alpha}^{(1)} f(x) - \frac{1}{\Gamma(\alpha)} \int_{r_f(x)}^{\infty} (\log t)^{\alpha-1} f^*(t) \frac{dt}{t}.$$

It is worth to mention here that these two almost optimal decompositions and the corresponding operators Ω are also valid for the pair (L^1, L^∞), (see example 10.6 above). Suppose that $T : L^1 \to L^{1,\infty}$ and $T : L^\infty \to L^\infty$, for instance let $T = P$, the Hardy operator. Then we can take the operators $\Omega^{(1)}$ either $\Omega^{(2)}$ for each one of the pairs (L^1, L^∞), $(L^{1,\infty}, L^\infty)$. Since

$$tf^*(t) \leq \sup_{0<s<t} sf^*(s) \leq \int_0^t f^*$$

it is clear that

$$(L^1, L^\infty)_{p,w,K} = (L^{1,\infty}, L^\infty)_{p,w,K} = \Lambda(p, w)$$

(cf. 4.1) and so, if T is a bounded operator $T : L^1 \to L^{1,\infty}$, $T : L^\infty \to L^\infty$, $1 \leq p < \infty$, $w \in \mathcal{C}B_p$, there exists a constant $C > 0$ such that the following estimates are true

$$\|T(f \log r_f) - T(f) \log r_{Tf}\|_{\Lambda(p,w)} = \|T(\Omega^{(1)} f) - \Omega^{(1)}(Tf)\|_{\Lambda(p,w)} \leq C\|f\|_{\Lambda(p,w)},$$

$$\left\| \int_{r_{(Tf)}(x)}^{\infty} \frac{(Tf)^*(t)}{t} dt \right\|_{\Lambda(p,w)} = \|\Omega^{(2)}(Tf) - \Omega^{(1)}(Tf)\|_{\Lambda(p,w)}$$

$$\leq \|\Omega^{(2)}(Tf) - T(\Omega^{(1)} f)\|_{\Lambda(p,w)} + \|T(\Omega^{(1)} f) - \Omega^{(1)}(Tf)\|_{\Lambda(p,w)} \leq C\|f\|_{\Lambda(p,w)}$$

and

$$\left\| T\left(\int_{r_f(x)}^{\infty} \frac{f^*(t)}{t} dt \right) \right\|_{\Lambda(p,w)} = \|T(\Omega^{(1)} f) - T(\Omega^{(2)} f)\|_{\Lambda(p,w)}$$

$$\leq \|T(\Omega^{(1)} f) - \Omega^{(1)}(Tf)\|_{\Lambda(p,w)} + \|\Omega^{(1)}(Tf) - T(\Omega^{(2)} f)\|_{\Lambda(p,w)} \leq C\|f\|_{\Lambda(p,w)},$$

for $f \in \Lambda(p, w)$.

Similar results apply to the pair $(L^p, L^q), 0 < p < q \leq +\infty$. In fact, a possible choice for Ω in this case is once again

$$\Omega f(x) = \frac{-1}{\theta} f(x)(\log r_f(x))$$

with $\theta = 1/p - 1/q$.

Example 10.8.

We consider pairs of vector valued L^p spaces based on a fixed measure space, eg. $\mathbb{R}^n$. Let (X_0, X_1) be a given Banach pair, let $1 \leq p \leq \infty$, and consider the pair $(L^p(X_0), L^p(X_1))$. Note that (cf. [**CJMR**], page 212)

$$(D_K(t, L^p(X_0), L^p(X_1))f)(s) = D_K(t, X_0, X_1)(f(s)).$$

In fact, this is just a reformulation of a result due to Cwikel [**Cw1**], we give the details for the convenience of the reader. Observe that if $f = D_K(t, L^p(X_0), L^p(X_1))f + (I - D_K(t, L^p(X_0), L^p(X_1)))f$, then

$$K(t, f; L^p(X_0), L^p(X_1))$$

$$\sim \|D_K(t, L^p(X_0), L^p(X_1))f\|_{L^p(X_0)} + \|(I - D_K(t, L^p(X_0), L^p(X_1)))f\|_{L^p(X_1)}$$

and therefore for each $s \in M$ (the ambient measure space) we shall have

$$f(s) = \left(D_K(t, L^p(X_0), L^p(X_1))f\right)(s) + \left((I - D_K(t, L^p(X_0), L^p(X_1)))f\right)(s)$$

and

$$K(t, f(s), X_0, X_1)$$
$$\leq \left\|\left(D_K(t, L^p(X_0), L^p(X_1))f\right)(s)\right\|_{X_0} + t\left\|\left((I - D_K(t, L^p(X_0), L^p(X_1)))f\right)(s)\right\|_{X_1}.$$

Consequently,

$$\left(\int_M K(t, f(s), X_0, X_1)^p ds\right)^{1/p}$$
$$\leq \|D_K(t, L^p(X_0), L^p(X_1))f(.)\|_{L^p(X_0)} + t\,\|(I - D_K(t, L^p(X_0), L^p(X_1)))f(.)\|_{L^p(X_1)}$$
$$\sim K(t, f; L^p(X_0), L^p(X_1)).$$

On the other hand, if $f(s) = D_K(t, X_0, X_1)f(s) + (I - D_K(t, X_0, X_1))f(s)$, then we have

$$K(f(s), X_0, X_1) \simeq \|D_K(t, X_0, X_1)f(s)\|_{X_0} + \|(I - D_K(t, X_0, X_1))f(s)\|_{X_1}$$

and therefore

$$C_p\left(\int_M K(t, f(s), X_0, X_1)^p ds\right)^{1/p}$$
$$\geq \left\|\|D_K(t, X_0, X_1)f(s)\|_{X_0}\right\|_{L^p} + \left\|\|(I - D_K(t, X_0, X_1))f(s)\|_{X_1}\right\|_{L^p}$$
$$\geq CK(t, f; L^p(X_0), L^p(X_1)).$$

It follows that

$$\left(\int_M K(t, f(s), X_0, X_1)^p ds\right)^{1/p} \sim K(t, f; L^p(X_0), L^p(X_1))$$

and that a possible choice for optimal decompositions can be obtained by the formulae

$$\left(D_K(t, L^p(X_0), L^p(X_1))f\right)(.) = D_K(t, X_0, X_1)f(.),$$
$$\left((I - D_K(t, L^p(X_0), L^p(X_1)))f\right)(.) = (I - D_K(t, X_0, X_1))f(.).$$

Thus, formally we shall have

$$\left(\Omega_{(L^p(X_0), L^p(X_1))}f\right)(s)$$
$$= \int_0^1 \left(D_K(t, L^p(X_0), L^p(X_1))f\right)(s)\frac{dt}{t} - \int_1^\infty \left((I - D_K(t, L^p(X_0), L^p(X_1)))f\right)(s)\frac{dt}{t}$$
$$= \int_0^1 D_K(t, X_0, X_1)f(s)\frac{dt}{t} - \int_1^\infty (I - D_K(t, X_0, X_1))f(s)\frac{dt}{t}$$
$$= \Omega_{(X_0, X_1)}(f(s)).$$

For example, consider $X_0 = L^p(w_0), X_1 = L^p(w_1)$, then from

$$\left(D_K(t, L^\infty(L^p(w_0)), L^\infty(L^p(w_1))f\right)(s) = D_K(t, L^p(w_0), L^p(w_1))f(s)$$

it follows that

$$\Omega_{(L^\infty(L^p(w_0)), L^\infty(L^p(w_1)))}f(x, y) = f(x, y)\log\frac{w_0(y)}{w_1(y)}.$$

As pointed out in [**CJMR**] and more recently in [**CCS1**] these ideas have interesting applications. For example they can be used to extend many of the results obtained by Segovia and Torrea and other authors for commutators of vector valued singular integral operators (cf. [**SeT**]).

Example 10.9.

We shall now consider the pairs $(H^{p_0}(\mathbb{R}^n), H^{p_1}(\mathbb{R}^n)), 0 < p_0 < p_1 < \infty$. It will be convenient to recall here the notion of a retract (cf. also section 3 above). A pair $\bar{A}$ is said to be a retract of a pair $\bar{B}$ if there exist linear operators $U : \bar{A} \to \bar{B}$, and $\Pr : \bar{B} \to \bar{A}$, such that $\Pr \circ U = id$ on $\bar{A}$. Under these conditions the optimal decompositions $D_K(t, \bar{A})f$ are given by

$$D_K(t, \bar{A})f = \Pr(D_K(t, \bar{B})(U(f))$$

(cf. [**JRW**]). Let us also recall that we can regard $(H^{p_0}(\mathbb{R}^n), H^{p_1}(\mathbb{R}^n))$ as a retract of the pair $(T_2^{p_0}(\mathbb{R}_+^{n+1}), T_2^{p_1}(\mathbb{R}_+^{n+1}))$ of Tent spaces (cf. [**CMS**], [**AM2**], and also Example 6.2 above). In fact in this case the operators in question are given by

$$U(f)(x,t) = tu_f(x,t), \ \Pr(f)(x) = \pi_\phi(f)(x)$$

where $u_f(x,t)$ is the Poisson integral of f, and

$$\pi_\phi(f) = \int_0^\infty f(\cdot, t) * \phi_t \frac{dt}{t},$$

where ϕ satisfies sufficient moment cancellation conditions and moreover

$$-2\pi \int_0^\infty \hat{\phi}(\xi t)\, |\xi|\, e^{-2\pi|\xi|t} dt = 1.$$

It follows that the optimal decompositions for the pair $(H^{p_0}(\mathbb{R}^n), H^{p_1}(\mathbb{R}^n))$ are then given by

$$D_K(t, (H^{p_0}(\mathbb{R}^n), H^{p_1}(\mathbb{R}^n))f = \pi_\phi(D_K(t, T_2^{p_0}(\mathbb{R}_+^{n+1}), T_2^{p_1}(\mathbb{R}_+^{n+1}))(U(f)).$$

Moreover, we have

$$K(t, f; T_2^{p_0}, T_2^{p_1}) \sim K(t, A_2(f), L^{p_0}(\mathbb{R}^n), L^{p_1}(\mathbb{R}^n))$$

and, with $1/p = 1/p_0 - 1/p_1$, we can select

$$D_K(t, T_2^{p_0}(\mathbb{R}_+^{n+1}), T_2^{p_1}(\mathbb{R}_+^{n+1}))f = f\chi_{T\{x: A_2(f)(x) > A_2(f)^*(t^p)\}}$$

where for a given set H on $\mathbb{R}^n$, $T(H)$ denotes 'the Tent with base H' defined by

$$T(H) = \{(x,t) \in \mathbb{R}_+^{n+1} : B(x,t) \subset H\},$$

and $B(x,t)$ denotes the ball with center x and radius t. It follows that

$$\Omega_{(T_2^{p_0}, T_2^{p_1})} f(x,s) = \int_0^1 f(x,s)\chi_{T_t(f)}(x,s))\frac{dt}{t} - \int_1^\infty (f(x,s) - f\chi_{T_t(f)}(x,s))\frac{dt}{t}.$$

Consequently, if we let $T_t(g) = T(\{x : A_2(f)(x) > A_2(g)^*(t^p)\})$

$$D_K(t, (H^{p_0}(\mathbb{R}^n), H^{p_1}(\mathbb{R}^n))f = \pi_\phi(U(f)\chi_{T_t(U(f))})$$

and

$$\Omega_{(H^{p_0}(R^n), H^{p_1}(R^n))} f = \int_0^1 \pi_\phi(U(f)\chi_{T_t(U(f))})\frac{dt}{t} - \int_1^\infty \pi_\phi(U(f) - U(f)\chi_{T_t(U(f))})\frac{dt}{t}$$

$$\Omega_{(H^{p_0}(\mathbb{R}^n), H^{p_1}(\mathbb{R}^n))} f = \pi_\phi(\Omega_{(T_2^{p_0}, T_2^{p_1})} U(f)).$$

Example 10.10.

We consider the pair $(H^p(\mathbb{R}^n), \dot{W}_p^\alpha(\mathbb{R}^n))$. It is shown in [**Lu**] that we can choose

$$D_K(t)f = f - W_{t^{1/\alpha}}^\alpha f$$
$$(I - D_K(t))f = W_{t^{1/\alpha}}^\alpha f$$

with $(W_t^\alpha f)\hat{}(\xi) = e^{-|t\xi|^\alpha}\hat{f}(\xi)$, $\alpha > 0$, is the Abel-Poisson operator. Other choices of homogeneous multipliers are also possible. For example, we can also take

$$D_K(t)f = f - B_{t^{1/\alpha}}^{\alpha,\delta} f$$

where $\left(B_t^{\alpha,\delta}(t)f\right)\hat{}(\xi) = (1 - |t\xi|^\alpha)_+^\delta \hat{f}(\xi)$, are the Bochner Riesz operators. In this case we have

$$\Omega_{(H^p(\mathbb{R}^n), \dot{W}_p^\alpha(\mathbb{R}^n))} f = \Omega f = \int_0^1 (f - W_{t^{1/\alpha}}^\alpha f)\frac{dt}{t} - \int_1^\infty W_{t^{1/\alpha}}^\alpha f \frac{dt}{t}$$

which formally gives

$$\begin{aligned}(\Omega f)\hat{}(\xi) &= \int_0^1 (1 - e^{-t|\xi|^\alpha})\hat{f}(\xi)\frac{dt}{t} - \int_1^\infty e^{-t|\xi|^\alpha}\hat{f}(\xi)\frac{dt}{t} \\ &= \left(\int_0^1 (1 - e^{-t|\xi|^\alpha})\frac{dt}{t} - \int_1^\infty e^{-t|\xi|^\alpha}\frac{dt}{t}\right)\hat{f}(\xi).\end{aligned}$$

Example 10.11.

The computation of nearly optimal decompositions entails the solution of a variational problem. When there is enough structure these problems turn out to be related to some classical problems in PDE's. We illustrate this point with some computations in the theory of Dirichlet spaces of Beurling and Denny (cf. [**BD**]). Let λ be a positive Radon measure on a locally compact Hausdorff space X, let $L_{loc}^1 = L_{loc}^1(X, d\lambda)$. A Hilbert space $D = D(X, \lambda)$ of complex valued functions $D \subset L_{loc}^1$, is called a Dirichlet space if it satisfies the following three axioms:

i) For each compact set $K \subset X$, there exists $c(K) > 0$ such that

$$\int_K |u|\, d\lambda \leq c(K)\, \|u\|_D\, .$$

ii) Let $C_0 = C_0(X)$ denote the space of complex valued continuous functions with compact support, then it holds that

$$C_0 \cap D \text{ is dense in } C_0 \text{ and } D.$$

iii) A normalized contraction in the complex plane $\mathbb{C}$ is a map $T : \mathbb{C} \to \mathbb{C}$, such that $|Tx - Ty| \leq |x - y|$, for all $x, y \in \mathbb{C}$, and moreover $T(0) = 0$. If $u \in D$, and T is a normalized contraction of $\mathbb{C}$, define $Tu(x) = T(u(x))$, then

$$\|Tu\|_D \leq \|u\|_D\, .$$

This last property is equivalent to the following statement: if $u \in D$, and v is such that

$$|v(x) - v(y)| \leq |u(x) - u(y)|$$
$$|v(x)| \leq |u(x)|$$

for all $x, y \in X$, then it follows that $v \in D$, and

$$\|v\|_D \leq \|u\|_D \, .$$

Consider the pair (L^2, D), then given $f \in L^2$ let

$$K_2(t, f; L^2, D)^2 = \inf\{\|f - u\|_2^2 + t\,\|u\|_D^2 : u \in D\}.$$

The following facts are shown in [**BD**]. It is observed that there exists a unique $u_t = (I - D_{K_2}(t))f \in D$ that minimizes the quadratic functional K_2 defined above. Moreover, this minimizer satisfies the variational condition :$\forall\ v \in L^2 \cap D$, it holds

$$t\langle (I - D_{K_2}(t))f, v\rangle_D - \int (D_{K_2}(t)f)\bar{v}d\lambda = 0. \tag{10.1}$$

For each $t > 0$, the map $f \to (I - D_{K_2}(t))f$ can be defined on D and is a linear and bounded contraction $(I - D_{K_2}(t)) : L^2 \to L^2$, and $(I - D_{K_2}(t)) : D \to D$. If $\|(I - D_{K_2}(t))f\|_D = \|f\|_D$ and $f \in D$, then $f = 0$. For each f bounded with compact support there exists a unique u_f such that for all $u \in D$, we have

$$(u_f, u)_D = \int f\bar{u}d\lambda.$$

Then u_f is denoted by Δf and is called the potential generated by f. More generally one can define potentials associated with measures in a similar fashion. Potentials associated with positive measures are called 'pure potentials.' Therefore we see that for a given $f \in D$, $t > 0$, $(I - D_{K_2}(t))f$ is thus a solution of the equation

$$u + t\Delta u = f.$$

In fact, let $v \in D$, then by the definition of potential and (10.1) we get

$$\begin{aligned}
&\int ((I - D_{K_2}(t))f + t\Delta(I - D_{K_2}(t))f)\,\bar{v}d\lambda \\
&= \int (I - D_{K_2}(t))f\bar{v}d\lambda + t\int \Delta(I - D_{K_2}(t))f\bar{v}d\lambda \\
&= \int (I - D_{K_2}(t))f\bar{v}d\lambda + t\langle (I - D_{K_2}(t))f, \bar{v}\rangle_D \\
&= \int (I - D_{K_2}(t))f\bar{v}d\lambda + \int (D_{K_2}(t)f)\bar{v}d\lambda \\
&= \int f\bar{v}d\lambda.
\end{aligned}$$

Using (10.1) and the fact that $D_{K_2}(t)f \to f$ we have that for each $f \in D, v \in D$

$$\begin{aligned}
\lim_{t\to\infty} t\langle (I - D_{K_2}(t))f, v\rangle_D &= \lim_{t\to\infty} \int (D_{K_2}(t)f)\bar{v}d\lambda \\
&= \int f\bar{v}d\lambda
\end{aligned}$$

and therefore we see that

$$\lim_{t\to\infty} t(I - D_{K_2}(t))f = \Delta f, \text{ in } D.$$

These remarks show that there are considerable structural relations between the computation of optimal decompositions and the solution of partial differential equations. We hope to develop this point further elsewhere.

11. BMO type spaces associated to Calderón weights

Originally the motivation that led John and Nirenberg to introduce and study the space BMO($\mathbb{R}$) were some problems in PDE's. The fundamental bridge with harmonic analysis was later provided by Fefferman's characterization of BMO($\mathbb{R}$) as the dual of $H^1(\mathbb{R})$. In classical analysis and interpolation theory BMO($\mathbb{R}$) appears as a limiting space for the scale of L^p spaces. Moreover, BMO($\mathbb{R}$) is also deeply related with the theory of weighted norm inequalities for the classical operators in analysis. Indeed, the functions in BMO($\mathbb{R}$) are multiple of logarithms of the weights in the class $\mathcal{A}_\infty$, which is the union of the $\mathcal{A}_p$ Muckenhoupt classes for $p \geq 1$. In still another context BMO($\mathbb{R}$) is exactly the class of functions ϕ for which the commutator $[H, M_\phi]$ is bounded in L^2, where H is the Hilbert transform and M_ϕ is the operator multiplication by the function ϕ. A good reference for the development of these interconnections is [**GR**], which also contains an extensive bibliography.

In view of this background, in this part of the paper we shall explore the analogs of a theory of functions of bounded mean oscillation in the context of our previous developments. Following through the connection between the classical space BMO($\mathbb{R}$) and the Muckenhoupt $\mathcal{A}_p$ classes of weights led us to consider spaces of bounded mean oscillation associated with the Calderón $\mathcal{C}_p$ classes of weights. In this context it is natural to consider BMO$(0,\infty)$ type spaces which are roughly speaking the set of multiples of logarithms of weights in $\mathcal{C}_p$. Let us informally denote this class by BCO_1 ("bounded Calderón oscillation"), we shall give formal definitions below. In analogy with the classical theory we should consider commutators $[S, M_\phi]$ where we replace the Hilbert transform H by the Calderón operator S and let $\phi \in \mathrm{BCO}_1$. The methods of previous sections lead us to the following result

PROPOSITION 11.1. *Let w_0, w_1 be two weights in $\mathcal{C}_p$, $1 \leq p < \infty$ and assume that $0 < \theta < 1$, then the commutator $[S, M_{\log(w_1 w_0^{-1})}]$ is bounded from $L^p(w_0^{1-\theta} w_1^\theta)$ into $L^p(w_0^{1-\theta} w_1^\theta)$.*

Furthermore, if $w \in \mathcal{C}_2$, then, for all $2 \leq p < \infty$, we have $[S, M_{\log w}] : L^p \to L^p$ and $\|[S, M_{\log w}]\|_{p \to p} \leq Cp^2 \|w\|_{\mathcal{C}_2}$.

PROOF. Let us consider the first half of the proposition. We have $S : L^p(w_i) \to L^p(w_i)$, $i = 0, 1$, and moreover that the operator Ω associated with the pair $(L^p(w_0), L^p(w_1))$ is given by $\Omega = M_{\log(w_1 w_0^{-1})}$. Then, the first part of the proposition follows by interpolation using the commutator theorem of [**JRW**], since

$$L^p(w_0^{1-\theta} w_1^\theta) = (L^p(w_0), L^p(w_1))_{\theta,p;K}.$$

To prove the second half let us remark that $w \in \mathcal{C}_2$ if and only if $w^{-1} \in \mathcal{C}_2$ and besides $\mathcal{C}_2 \subseteq \mathcal{C}_p$. Also we note that the corresponding operator Ω associated to the pair $(L^p(w), L^p(w^{-1}))$ is given by $\Omega f = \frac{C}{p} f \log w$. Hence, by interpolation and Proposition 2.9, we arrive at

$$\|[S, M_{\log w}]\|_{p\to p} = Cp\|[S,\Omega]\|_{p\to p} \le Cp\|w\|_{\mathcal{C}_p} \le Cp^2\|w\|_{\mathcal{C}_2}.$$

□

The next proposition ensures that if $w \in \mathcal{C}_2$ then the oscillation of $\log w$ is under control.

LEMMA 11.2. *Let ϕ be a measurable function on $(0,+\infty)$. Let $w = e^\phi$ be a weight. Then $w \in \mathcal{C}_2$ if and only if there exists a constant C such that for all $t > 0$,*

$$\frac{1}{t}\int_0^t \exp\left|\phi(x) - t\int_t^\infty \phi(y)\frac{dy}{y^2}\right| dx \le C \tag{11.1}$$

$$t\int_t^\infty \exp\left|\phi(x) - \frac{1}{t}\int_0^t \phi(y)dy\right|\frac{dx}{x^2} \le C. \tag{11.2}$$

PROOF. If $w = e^\phi \in M^2$ then

$$\frac{1}{t}\int_0^t \exp\left(\phi(x) - t\int_t^\infty \phi(y)\frac{dy}{y^2}\right)dx = \frac{1}{t}\exp\left(-t\int_t^\infty \phi(y)\frac{dy}{y^2}\right)\int_0^t \exp(\phi(x))dx \le$$

$$\le \frac{1}{t}t\left(\int_t^\infty \exp(-\phi(x))\frac{dx}{x^2}\right)\left(\int_0^t \exp(\phi(x))dx\right) \le \|w\|_{M^2},$$

where we have used Jensen's inequality with e^{-x} and the probability measure $d\mu = \frac{tdx}{x^2}$ on $[t,+\infty)$ and the M^2 condition for w.

If $w = e^\phi \in M_2$, we have

$$t\int_t^\infty \exp\left(\phi(x) - \frac{1}{t}\int_0^t \phi(y)dy\right)\frac{dx}{x^2} = t\exp\left(-\frac{1}{t}\int_0^t \phi(y)dy\right)\int_t^\infty \exp(\phi(x))\frac{dx}{x^2} \le$$

$$\le t\frac{1}{t}\left(\int_0^t \exp(-\phi(x))dx\right)\left(\int_t^\infty \exp(\phi(x))\frac{dx}{x^2}\right) \le \|w\|_{M_2}.$$

Since $e^\phi \in \mathcal{C}_2$ if and only if $e^{-\phi} \in \mathcal{C}_2$ we also get (11.1) and (11.2).

Conversely, if (11.1) and (11.2) are fulfilled, then

$$\int_t^\infty \exp(\phi(x))\frac{dx}{x^2} =$$

$$= \left(\int_t^\infty \exp\left(\phi(x) - \frac{1}{t}\int_0^t \phi(y)dy\right)\frac{dx}{x^2}\right)\exp\left(\frac{1}{t}\int_0^t \phi(x)dx\right) \le C\exp\left(\frac{1}{t}\int_0^t \phi(x)dx\right)$$

and

$$\int_0^t \exp(-\phi(x))dx =$$

$$= \left(\int_0^t \exp\left(-\phi(x) + t\int_t^\infty \phi(y)\frac{dy}{y^2}\right)dx\right)\exp\left(-t\int_t^\infty \phi(y)\frac{dy}{y^2}\right)$$

$$\le C\exp\left(-t\int_t^\infty \phi(y)\frac{dy}{y^2}\right).$$

Therefore,

$$\left(\int_t^\infty \exp(\phi(x))\frac{dx}{x^2}\right)\left(\int_0^t \exp(-\phi(x))dx\right) \le C\exp\left(\frac{1}{t}\int_0^t\left(\phi(y)-t\int_t^\infty \phi(x)\frac{dx}{x^2}\right)dy\right)$$

$$\le \frac{C}{t}\int_0^t \exp\left(\phi(y)-t\int_t^\infty \phi(x)\frac{dx}{x^2}\right)dy \le C$$

which is the condition M_2.

The condition M^2 can be obtained using a similar argument. □

In the sequel we shall use a probabilistic notation. E_t will represent the expectation in the probability space $((0,t), dx/t)$ and E^t the corresponding expectation in the probability space $((t,\infty), tdx/x^2)$.

If we are careful with the constants appearing in the proof of the preceding proposition, what we have achieved is that: if for a measurable function ϕ on $(0,+\infty)$, we have $e^\phi \in \mathcal{C}_2$ then, for all $t>0$, it holds

$$E_t \exp|\phi - E^t\phi| \le C_1$$

$$E^t \exp|\phi - E_t\phi| \le C_2$$

with $\max\{C_1, C_2\} \le \|w\|_{M_2}^2 + \|w\|_{M^2}^2$. Conversely, if a function ϕ satisfies, for some constants $C_1, C_2 \ge 1$, and all $t>0$

$$E_t \exp|\phi - E^t\phi| \le C_1$$

$$E^t \exp|\phi - E_t\phi| \le C_2$$

then $w = e^\phi \in \mathcal{C}_2$ with

$$\|w\|_{M_2} \le C_1 C_2^{1/2} \le C_1C_2$$

$$\|w\|_{M^2} \le C_1^{1/2} C_2 \le C_1C_2.$$

The previous discussion motivates the following formal definition of BCO_1.

DEFINITION 11.3. *We say that a measurable function ϕ belong to the class BCO_1 if there exist constants $\lambda>0$ and $C>0$ such that*

$$E_t \exp(\lambda|\phi - E^t\phi|) + E^t \exp(\lambda|\phi - E_t\phi|) \le C$$

for all $t>0$.

Lemma 11.2 ensures that if $w = e^\phi \in \mathcal{C}_2$ then $\log w = \phi \in \mathrm{BCO}_1$ and reciprocally if $\phi \in \mathrm{BCO}_1$, then there exists $\lambda > 0$ such that $e^{\lambda\phi} \in \mathcal{C}_2$. It is also obvious that if $\phi \in \mathrm{BCO}_1$, then there exists $\lambda>0$ such that $e^{\mu\phi}$ is in $\mathcal{C}_2$, for all $|\mu| \le \lambda$. Therefore,

$$\mathrm{BCO}_1 = \{\lambda \log w : \lambda \in \mathbb{R}, w \in \mathcal{C}_2\}.$$

Actually, the same is true for $1<p<\infty$.

PROPOSITION 11.4. *Given p, $1<p<\infty$, we have*

$$BCO_1 = \{\lambda \log w : \lambda \in \mathbb{R}, w \in \mathcal{C}_p\}.$$

PROOF. If $p \leq 2$, $w \in \mathcal{C}_p \subset \mathcal{C}_2$ implies that $\log w \in \mathrm{BCO}_1$. On the other hand, suppose that $1 < p < 2$. If $\phi \in \mathrm{BCO}_1$, then $w_0 = e^{\lambda\phi} \in \mathcal{C}_2$, for some $\lambda > 0$. Then there exists $\alpha > 0$ such that $w^\alpha \in \mathcal{C}_p$. Indeed, from the note that follows Proposition 3.6 above, we have that $w = w_0^{p(1-\theta)/p_0} w_1^{p\theta/p_1} \in \mathcal{C}_p$ whenever $w_0 \in \mathcal{C}_{p_0}$, $w_1 \in \mathcal{C}_{p_1}$, $\dfrac{1}{p} = \dfrac{1-\theta}{p_0} + \dfrac{\theta}{p_1}$. If we take $p_0 = 2$, and select $p_1 < p$ near to 1, $w_0 = w$, $w_1 = 1$ and θ such that $\dfrac{1}{p} = \dfrac{1-\theta}{p_0} + \dfrac{\theta}{p_1}$, then $\alpha = p(1-\theta)/2$.

The case $p \geq 2$ is a consequence of preceding situation since it is enough to observe that $w^{-p'/p} \in \mathcal{C}_{p'} \subset \mathcal{C}_2$. □

We cannot expect a perfect similarity between the class BCO_1, we have just introduced, and the classical space BMO. In fact, an inequality in the John-Nirenberg' style is not true. For instance, the function $\phi(x) = |\log(x-1)|^2 \chi_{[1,2]}(x)$ is not in BCO_1 but it satisfies the following, given $1 \leq p < \infty$ there is a constant $C_p > 0$ such that

$$E_t\left|\phi - E^t\phi\right|^p \leq C_p$$

and

$$E^t\left|\phi - E_t\phi\right|^p \leq C_p.$$

for all $t > 0$.

Our next result can be seen as a weak form of the John-Nirenberg lemma. Given a measurable function ϕ on $(0, +\infty)$, we denote $\phi^*(x) = \phi(1/x)$. Remark that a simple change of variables shows that $\phi \in \mathrm{BCO}_1$ if and only if $\phi^* \in \mathrm{BCO}_1$ with the same constants and that

$$E^t \exp|\phi - E_t\phi| = E_{1/t}\exp|\phi^* - E^{1/t}\phi^*|.$$

LEMMA 11.5. *$\phi \in BCO_1$ if and only if there exists a constant $C \geq 1$ such that for every $1 \leq p < \infty$ and for all $t > 0$ we have*

$$\begin{aligned} \left(E_t|\phi - E^t\phi|^p\right)^{1/p} &\leq Cp \\ \left(E_t|\phi^* - E^t\phi^*|^p\right)^{1/p} &\leq Cp. \end{aligned} \tag{11.3}$$

PROOF. If $\phi \in \mathrm{BCO}_1$, then there exist two positive constants C, λ, such that for all $t > 0$

$$E_t \exp(\lambda|\phi - E^t\phi|) \leq C$$

$$E_t \exp(\lambda|\phi^* - E^t\phi^*|) \leq C.$$

Then, by expanding the exponential function in Taylor series, we have that, for all $n \in \mathbb{N}$ and for all $t > 0$,

$$E_t|\phi - E^t\phi|^n \leq C\frac{n!}{\lambda^n}$$

$$E_t|\phi^* - E^t\phi^*|^n \leq C\frac{n!}{\lambda^n}.$$

Thus, we have obtained the result for all $n \in \mathbb{N}$ and by interpolation for all $p \geq 1$.

For the converse, if we assume that (11.3) is true, in particular we have

$$E_t|\phi - E^t\phi|^n \leq (Cn)^n$$

$$E_t|\phi^* - E^t\phi^*|^n \leq (Cn)^n$$

for all $n = 0, 1, 2, \dots$ and for all $t > 0$. If we pick $\lambda > 0$ such that $\lambda Ce < 1$ then multiplying by $\lambda^n/n!$ in the last inequalities and summing for $n = 0, 1, 2, ..,$ we arrive at

$$E_t \exp(\lambda|\phi - E^t\phi|) \leq \sum_{n=0}^{\infty} \frac{(Cn\lambda)^n}{n!} \leq C$$

which is one half of the condition (11.1). Since the same argument applies to ϕ^* the result follows. □

REMARK.

We should note that, as a consequence of the commutator theorem (cf. Proposition 11.1), we have that for $\phi \in \mathrm{BCO}_1$

$$\|[S, M_\phi]\|_{p\to p} \sim \|[S, M_{\phi^*}]\|_{p\to p} = O(p^2), \ (p \to \infty). \tag{11.4}$$

In particular, if f is any measurable function, then

$$[S, M_\phi]f(x) = \int_0^\infty \min\left\{\frac{1}{x}, \frac{1}{y}\right\}(\phi(y) - \phi(x))f(y)dy$$

and therefore applying (11.4) to ϕ, ϕ^* and $f(x) = x^{-1}\chi_{[t,\infty]}(x)$, we get that, for all $1 < p < \infty$,

$$\begin{aligned} E_t|\phi - E^t\phi|^p &= t^{p-1}\int_0^t \left|\int_t^\infty (\phi(x) - \phi(y))\frac{dy}{y^2}\right|^p \\ &\leq t^{p-1}\int_0^\infty \left|\int_t^\infty \min\{x^{-1}, y^{-1}\}(\phi(x) - \phi(y))\frac{dy}{y}\right|^p \\ &= t^{p-1}\|[S, M_\phi]f\|_p^p \leq t^{p-1}(Cp^2)^p\|f\|_p^p = \frac{(Cp^2)^p}{p-1} \end{aligned}$$

The same computations can be also applied to the function ϕ^*, and hence we obtain a weaker version of (11.3). We don't know if (11.4) can be improved to

$$\|[S, M_\phi]\|_{p\to p} \sim \|[S, M_{\phi^*}]\|_{p\to p} = O(p). \tag{11.5}$$

If we consider the operator P, then Proposition 2.9 and the commutator theorem imply that

$$\|[P, M_\phi]\|_{p\to p} \sim \|[P, M_{\phi^*}]\|_{p\to p} = O(p). \tag{11.6}$$

For locally integrable functions on $(0, \infty)$ we define an adapted "maximal sharp function" (cf. [**MS**]) by

$$\phi^\#(x) = \phi(x) - \frac{1}{x}\int_0^x \phi(y)dy = \phi(x) - P\phi(x).$$

The role of this operator is illustrated in the following result

PROPOSITION 11.6. *For any function ϕ in BCO_1 we can find two positive constants $C, \lambda > 0$ such that*

$$E_t \exp|\lambda\phi^\#| \leq C$$

for all $t > 0$.

PROOF. By (11.6) there exists a constant $C > 0$ such that for all $p \geq 1$, and for all $f \in L^p$, we have

$$\|[P, M_\phi]f\|_p^p \leq C^p p^p \|f\|_p^p.$$

In particular, if we take $f = \chi_{[0,t]}$, then for any $t > 0$, we obtain

$$\int_0^\infty \left| \frac{1}{x} \int_0^x (\phi(y) - \phi(x))\chi_{[0,t]}(y)dy \right|^p dx \leq C^p p^p t.$$

Hence

$$E_t|\phi - P\phi|^p \leq C^p p^p.$$

At this point we can follow the argument of the second half of the proof of lemma 11.5 to achieve the required inequality for some $\lambda > 0$ and for all $t > 0$. □

NOTE.

Since $\phi \in \mathrm{BCO}_1$ if and only if $\phi^* \in \mathrm{BCO}_1$, we get

$$E^t(\lambda|\phi - \bar{P}\phi|) \leq C$$

for all $t > 0$, where

$$\bar{P}\phi(x) = x \int_x^\infty \frac{\phi(y)}{y^2} dy.$$

It is well known that weights in $\mathcal{A}_\infty$ can be characterized as having equivalent arithmetical and geometrical means over all cubes. In our situation we have a weaker result: the means we consider are computed over our probability spaces $(0,t), dx/t)$ and $(t,\infty), tdx/x^2)$, for all $t > 0$. To facilitate the discussion let us introduce a new class of weights

DEFINITION 11.7. *A weight w belongs to the class $\mathcal{C}_\infty^*$ if there exists a positive constant $C > 0$ such that, for all $t > 0$,*

$$\exp E_t \log w \leq E_t w \leq C \exp E_t \log w$$
$$\exp E^t \log w \leq E^t w \leq C \exp E^t \log w$$

Observe that the leftmost inequalities are always true by Jensen's inequality. Notice also that if $w = e^\phi$, then $w \in \mathcal{C}_\infty^*$ if and only if

$$E_t \exp|\phi - E_t\phi| \leq C$$
$$E^t \exp|\phi - E^t\phi| \leq C$$

for some constant $C > 0$ and for all $t > 0$.

PROPOSITION 11.8. *Let w be a $\mathcal{C}_p$ weight.*
i) If $1 \leq p \leq 2$ then $w^\lambda \in \mathcal{C}_\infty^*$, for all $|\lambda| \leq 1/2$
ii) If $2 \leq p < \infty$ then $w^\lambda \in \mathcal{C}_\infty^*$, for all $|\lambda| \leq p'/2p$.

PROOF. We only prove the required inequalities for E_t, the corresponding ones for E^t can be obtained in a similar fashion. Suppose that $w = e^\phi \in \mathcal{C}_2$, then, by Lemma 11.2, we have

$$E_t \exp|\phi - E^t\phi| \leq C$$

for all $t > 0$. Therefore,

$$\begin{aligned}
E_t \exp(\lambda^n |\phi - E_t\phi|) &= \sum_{n=0}^{\infty} \frac{\lambda^n}{n!} E_t|\phi - E^t\phi|^n \\
&= \sum_{n=0}^{\infty} \frac{\lambda^n}{n!} \left(\sup_{\|g\|_{n'} \le 1} |E_t(\phi - E_t\phi)g| \right)^n \\
&= \sum_{n=0}^{\infty} \frac{\lambda^n}{n!} \left(\sup_{\|g\|_{n'} \le 1} |E_t\phi(g - E_t g)| \right)^n \\
&= \sum_{n=0}^{\infty} \frac{\lambda^n}{n!} \left(\sup_{\|g\|_{n'} \le 1} |E_t(\phi - E^t\phi)(g - E_t g)| \right)^n \\
&\le \sum_{n=0}^{\infty} \frac{\lambda^n}{n!} \left(\sup_{\|g\|_{n'} \le 1} \left(E_t|\phi - E^t\phi|^n\right)^{1/n} \left(E_t|g - E_t g|^n\right)^{1/n} \right)^n \\
&\le \sum_{n=0}^{\infty} \frac{(2\lambda)^n}{n!} E_t|\phi - E^t\phi|^n \\
&= E_t \exp(2\lambda|\phi - E^t\phi|).
\end{aligned}$$

If $1 \le p \le 2$, since $w \in \mathcal{C}_2$, we take $|\lambda| \le 1/2$ and if $2 \le p < \infty$, since $w^{-p'/p} \in \mathcal{C}_2$, we take $|\lambda| \le p'/2p$. Under these conditions we see that

$$E_t \exp|\lambda(\phi - E_t\phi)$$

12. Atomic decompositions and duality

In this section we consider the predual spaces associated with the spaces of bounded mean oscillation introduced in §11.

For the convenience of the reader let us now recall some facts concerning Orlicz spaces (for detailed proofs of these results we refer for instance to [**Ku**]).

An Orlicz function Φ is a continuous, non decreasing, non negative, convex function defined on $[0, \infty)$ with $\Phi(0) = 0$ and $\lim_{t\to 0} \Phi(t) = \infty$.

Given an Orlicz function Φ and a probability space (Ω, μ) the corresponding Orlicz space $L^\Phi(\Omega)$ is the function space of all measurable functions f defined on Ω for which there exists a $\lambda > 0$ such that

$$\int_\Omega \Phi(\lambda|f|)d\mu < \infty.$$

We consider the Luxemburg norm on Orlicz spaces, defined by

$$\|f\|_\Phi = \inf\{\rho > 0; \int_\Omega \Phi\left(\frac{1}{\rho}|f|\right) d\mu \leq 1\}.$$

The space $L^\Phi(\Omega)$ is a Banach space endowed with this norm.

If Φ and Ψ are complementary Young functions they satisfy the inequality

$$uv \leq \Phi(u) + \Psi(v) \;\; u, v > 0.$$

Hölder 's inequality takes the form

$$\int_\Omega |fg|d\mu \leq 2\|f\|_\Phi\|g\|_\Psi.$$

Moreover, if Φ satisfies the Δ_2 condition at infinity then $L^\Psi(\Omega)$ is isomorphic to the dual space of $L^\Phi(\Omega)$. More precisely, if T is a bounded linear functional on $L^\Phi(\Omega)$ then there exists a uniquely determined g in $L^\Psi(\Omega)$ such that $Tf = \langle f, g\rangle$ and moreover

$$\frac{1}{2}\|g\|_\Psi \leq \|T\| \leq 2\|g\|_\Psi.$$

In our framework we shall only consider the pair of complementary Orlicz functions given by:

$$\Phi(t) = t\log^+ t, \;\; 0 \leq t$$

and

$$\Psi(t) = \begin{cases} t, & \text{if } 0 \leq t \leq 1 \\ e^{t-1}, & \text{if } 1 \leq t \end{cases}$$

It follows that Φ satisfies Δ_2 condition at infinity and Ψ doesn't.

Since $\Psi(x) \leq e^x \leq 1 + e\Psi(x)$ for all $x > 0$, and we work on probability spaces, we see that the class BCO_2 can be also defined as the class of measurable functions on $(0, \infty)$ for which there exists a constant $\lambda > 0$ such that

$$\sup_{t>0} \max \left\{ E_t \Psi \left(\lambda \left| \phi - E_t \phi \right| \right), E^t \Psi \left(\lambda \left| \phi - E^t \phi \right| \right) \right\} < \infty. \tag{12.1}$$

In the sequel we denote by $X_t = L^\Phi((0,t), dx/t)$, $X^t = L^\Phi((t,\infty), tdx/x^2)$, $Y_t = L^\Psi((0,t), dx/t)$ and $Y^t = L^\Psi((t,\infty), tdx/x^2)$.

The convexity of Ψ implies that, for all $0 < \lambda \leq 1$ and for all $x > 0$, we have $\Psi(\lambda x) \leq \lambda \Psi(x)$. Suppose that $\phi \in \mathrm{BCO}_2$, then, for all $t > 0$, we have $\phi - E_t \in Y_t$, $\phi - E^t \in Y^t$, and moreover, there exists a constant C such that

$$\|\phi - E_t\|_{Y_t} \leq C, \;\; \|\phi - E^t\|_{Y^t} \leq C.$$

Let

$$\begin{aligned} \|\phi\|_{\mathrm{BCO}_2} &= \sup_{t>0} \inf \left\{ \rho > 0; \max \left\{ E_t \Psi \left(\frac{1}{\rho} |\phi - E_t \phi| \right), E^t \Psi \left(\frac{1}{\rho} |\phi - E^t \phi| \right) \right\} \leq 1 \right\} \\ &= \sup_{t>0} \max \{ \|\phi - E_t \phi\|_{X_t}, \|\phi - E^t \phi\|_{X^t} \}. \end{aligned}$$

Convexity considerations show that the supremum exists and therefore, $\|\phi\|$ is well defined. It is also readily seen that $\|\phi\| = 0$ if and only if the function ϕ is a constant and that $\|\lambda \phi\| = |\lambda| \|\phi\|$ for all scalar λ and for all function $\phi \in \mathrm{BCO}_2$.

Let us sketch the proof of triangle inequality. Let ϕ_1, ϕ_2 two functions in BCO_2. For any $t > 0$,

$$\max \left\{ E_t \Psi \left(\frac{1}{\|\phi_i\|} |\phi - E_t \phi| \right), E^t \Psi \left(\frac{1}{\|\phi_i\|} |\phi - E^t \phi| \right) \right\} \leq 1$$

for $i = 1, 2$. By the convexity of Ψ, we get

$$\Psi \left(\frac{x_1 + x_2}{a_1 + a_2} \right) \leq \frac{a_1}{a_1 + a_2} \Psi\left(\frac{x_1}{a_1}\right) + \frac{a_2}{a_1 + a_2} \Psi\left(\frac{x_2}{a_2}\right)$$

and therefore we obtain

$$\|\phi_1 + \phi_2\| \leq \|\phi_1\| + \|\phi_2\|.$$

In this fashion the space BCO_2, modulo constants, which by abuse of notation we again denote by BCO_2, becomes a Banach space. This fact will be a by-product of our characterization of BCO_2 as the dual space of a corresponding atomic "like H^1" space.

DEFINITION 12.1. *A measurable function on $(0, \infty)$, a, is a C-atom if there exists a positive $t > 0$ such that the following conditions hold*

i) *a is supported in the interval $(0, t)$ (in which case we shall say it is a C_t-atom) or in the interval (t, ∞) (in which we shall say it is a C^t-atom),*

ii) $\int_0^\infty a(x) dx = 0$,

iii) $E_t \Phi(t|a|) \leq 1$, *if a is a C_t-atom and* $E^t \Phi(x^2 t^{-1} |a|) \leq 1$, *if a is a C^t-atom.*

Condition iii) means that if a is C_t-atom then it belongs to the Orlicz space X_t and $\|a\|_{X_t} \leq 1/t$. Likewise if a is C^t-atom then it belongs to $tx^{-2} X^t$ and $\|a\| = \|x^2 t^{-1} a\|_{X^t} \leq 1$.

In the next lemma we collect some properties of C-atoms.

LEMMA 12.2. *Let a be a C-atom, then*
i) $\|a\|_{L^1((0,\infty),dx)} \leq 1 + e$.

ii) *If a is a C_t-atom and is supported in the interval $(0, s)$ for some $s < t$ then a is a C_s-atom.*

iii) *If a is a C^t-atom and is supported in the interval (s, ∞) for some $s > t$ then a is also a C^s-atom.*

iv) *Let f be a measurable function supported in the interval (s, t), for some $0 < s < t$, such that $\int_0^\infty f(x)dx = 0$ and $\int_0^\infty \Phi(|f(x)|)dx < \infty$, then there is a constant $\lambda > 0$ such that λf is simultaneously a C_t-atom and a C^s-atom.*

v) *The function $a(x)$ is a C_t-atom if and only if $\bar{a}(x) = a(1/x)x^{-2}$ is a $C^{1/t}$-atom.*

PROOF. i) We use the fact that $x \leq \Phi(x)$, for $x \geq e$. Suppose that a is a C_t-atom, then

$$\int_0^\infty |a(x)|dx = (E_t|ta|\chi_{\{t|a|>e\}}) + (E_t|ta|\chi_{\{t|a|\leq e\}})$$
$$\leq E_t\Phi(t|a|) + e \leq 1 + e.$$

Suppose that a is a C^t-atom, then

$$\int_0^\infty |a(x)|dx = E^t(\frac{x^2|a|}{t}\chi_{\{x^2|a|>te\}}) + E^t(\frac{x^2|a|}{t}\chi_{\{x^2|a|\leq te\}})$$
$$\leq E^t\Phi(\frac{x^2|a|}{t}) + e \leq 1 + e.$$

The proof of ii) and iii) follows from the fact that, for convex functions, $\Phi(\lambda x) \leq \lambda\Phi(x)$ for all $0 < \lambda \leq 1$ and for all $x > 0$.

The assertion iv) is a consequence of the monotonicity of the function Φ and the monotone convergence theorem.

v) follows by a change of variables. □

Let us now introduce the space CH^1.

DEFINITION 12.3. *The class CH^1 consists of the functions f in $L^1((0, \infty), dx)$ which have an admissible representation $f = \sum_{n=1}^\infty \lambda_n a_n$, convergence in L^1, where the a_n's are C-atoms and $\sum_{n=1}^\infty |\lambda_n| < \infty$. Let*

$$\|f\|_{CH^1} = \inf \sum_{n=1}^\infty |\lambda_n|,$$

where the infimum runs over all admissible representations of f.

PROPOSITION 12.4. *CH^1 is a Banach space*

PROOF. Let $\mathcal{I}$ the class of all C-atoms. We consider the Banach space $\ell^1(\mathcal{I})$ of all summable families $(\lambda_a)_{a\in\mathcal{I}}$ with the natural ℓ^1-norm. The map from $\ell^1(\mathcal{I})$ into $L^1((0, \infty), dx)$ defined by

$$(\lambda_a)_{a\in\mathcal{I}} \longrightarrow \sum_{a\in\mathcal{I}} \lambda_a a$$

is a bounded linear operator. The kernel N of this operator is a closed subspace of $\ell^1(\mathcal{I})$ and it is very easy to see that CH^1 is isometrically isomorphic to the quotient space $\ell^1(\mathcal{I})/N$. □

LEMMA 12.5. *The following assertions are true:*

i) *If* $f \in CH^1$ *then*

$$\|f\|_{L^1((0,\infty),dx)} \leq (1+e)\|f\|_{CH^1}.$$

ii) *If* a *is a* C*-atom then* $\|a\|_{CH^1} \leq 1$.

iii) *The linear span of the* C*-atoms is dense in* CH^1.

iv) *A linear map defined on* CH^1 *and taking values in another normed space is continuous if and only if is uniformly bounded on the* C*-atoms.*

The proof is trivial.

We can now state and prove the following duality result

THEOREM 12.6. *The space* BCO_2 *is isomorphic to the dual space of* CH^1.

PROOF. First we prove that if ϕ is a function in BCO_2 then there exists a linear form T_ϕ on CH^1 such that $\|T_\phi\| \leq 2\|\phi\|_{\mathrm{BCO}_2}$.

Let a be C-atom. It is easy to see that $\int_0^\infty |a\phi| < \infty$, therefore the operator defined on C-atoms by

$$T_\phi(a) = \int_0^\infty a\phi$$

is well defined. If a is a C_t-atom, then by Hölder's inequality

$$\begin{aligned}|T_\phi(a)| &= |E_t(ta(\phi - E_t\phi))| \\ &\leq 2\|ta\|_{X_t}\|\phi - E_t\phi\|_{Y_t} \leq 2\|\phi\|_{\mathrm{BCO}_2}.\end{aligned}$$

Similarly if a is a C^t-atom then

$$\begin{aligned}|T_\phi(a)| &= \left|E^t(\frac{x^2a}{t}(\phi - E^t\phi))\right| \\ &\leq 2\|x^2t^{-1}a\|_{X^t}\|\phi - E^t\phi\|_{Y^t} \leq 2\|\phi\|_{\mathrm{BCO}_2}.\end{aligned}$$

Since the linear span generated by C-atoms is dense in CH^1 and T_ϕ is well defined there we find that T_ϕ is a bounded linear form on CH^1 defined by

$$T_\phi(f) = T_\phi\left(\sum_{n=1}^\infty \lambda_n a_n\right) = \sum_{n=1}^\infty \lambda_n \int_0^\infty a_n\phi$$

for any $f \in CH^1$. Moreover,

$$\|T_\phi\| \leq 2\|\phi\|_{\mathrm{BCO}_2}.$$

This concludes the first part of the proof.

Let T be an element in $(CH^1)^*$. Then, for any function $f \in CH^1$, $|T(f)| \leq \|T\|\|f\|_{CH^1}$.

Let $t > 0$ fixed. Consider the subspace $X_t(0) \subseteq X_t$ of functions f in X_t with $\int_0^t f = 0$. It is clear that $X_t(0)$ is also a subspace of CH^1, since $\lambda^{-1} f$ is a C_t-atom whenever $\lambda > t\|f\|_{X_t}$. So, $\|f\|_{CH^1} \leq t\|f\|_{X_t}$. Therefore, if we consider the restriction of T to the subspace $X_t(0)$ we have

$$|T(f)| \leq t\|T\|\|f\|_{X_t}.$$

Now, the Hahn-Banach theorem allows us to extend this functional to the whole space X_t without increasing its norm. Then, by duality, we see that this functional is given by a unique function $\phi \in Y_t$, with norm dominated by twice the norm of the functional. In this case, by considering the function ϕ/t and denoting it again by ϕ, we conclude that:

Claim 1. For any $t > 0$, there exists a function ϕ supported on the interval $(0, t)$ such that $\|\phi\|_{Y_t} \leq 2t\|T\|$ and if a is a C_t-atom, then $T(a) = \int_0^t a\phi$.

The function ϕ is unique modulo constants. Indeed, suppose we have two functions ϕ_1, ϕ_2 satisfying the claim, and let $\phi = \phi_1 - \phi_2$. Let f be any element in X_t, then, since the function $f - E_t f \in X_t(0)$ is a multiple of a C_t-atom, we have

$$\begin{aligned} 0 = \int_0^t (f - E_t f)\phi &= \int_0^t (f - E_t f)(\phi - E_t\phi) \\ &= \int_0^t f(\phi - E_t\phi). \end{aligned}$$

Thus $\phi - E_t\phi = 0$ and consequently ϕ is a constant.

By considering an increasing sequence of t's going to ∞ we can find a function ϕ defined in $(0, \infty)$ that satisfies $\|\phi\chi_{[0,t]}\|_{Y_t} \leq 2t\|T\|$ and if a is a C_t-atom, then $T(a) = \int_0^t a\phi$, for all $t > 0$.

Next we repeat the process with the intervals (t, ∞). We shall sketch the method of proof because some differences occur.

Let $t > 0$ fixed. Consider now the subspace $tx^{-2}X^t(0) \subseteq tx^{-2}X^t$ consisting of all functions f in $tx^{-2}X^t$ such that $\int_t^\infty f = 0$. It is clear that $tx^{-2}X^t(0)$ is also a subspace of CH^1, since $\lambda^{-1} f$ is a C^t-atom whenever $\lambda > \|t^{-1}x^2 f\|_{X^t}$. So, $\|f\|_{CH^1} \leq \|t^{-1}x^2 f\|_{X^t}$. If we consider the restriction of T to the subspace $tx^{-2}X^t$ we have

$$|T(f)| \leq \|T\|\|f\|_{tx^{-2}X^t}.$$

Now by the Hahn-Banach extension theorem we extend this functional to the whole space $tx^{-2}X^t$ and, by duality, this functional is represented by a unique function $\psi \in t^{-1}x^2Y^t$, with norm dominated by twice the norm of the functional. Proceding as before we conclude

Claim 2. For any $t > 0$, there exists a function ψ supported on the interval (t, ∞) such that $\|\psi\|_{Y^t} \leq 2t\|T\|$ and if a is a C^t-atom, then $T(a) = \int_t^\infty a\psi$.

This function is also unique modulo constants. For if ψ_1, ψ_2 are functions satisfying ii) consider $\psi = \psi_1 - \psi_2$. Let f be any element in $tx^{-2}X^t$, and select a function $f_0 \in tx^{-2}X^t$ such that $\int_t^\infty f_0 = 1$. Since the function $f - f_0 \int_0^\infty f \in tx^{-2}X^t(0)$ is a multiple of a C^t-atom, we have

$$\int_t^\infty f\psi = \left(\int_t^\infty f_0\psi\right)\left(\int_t^\infty f\right) = \int_t^\infty \left(\int_t^\infty f_0\psi\right) f$$

for any $f \in tx^{-2}X^t$. Thus $\psi = \int_t^\infty f_0\psi$ which is a constant.

We conclude in the same fashion as before and we obtain a function ψ defined on $(0,\infty)$ satisfying $\|\psi\chi_{[t,\infty]}\|_{Y^t} \le 2t\|T\|$ and if a is a C^t-atom, then $T(a) = \int_t^\infty a\psi$, for all $t > 0$.

We shall now prove that the functions ϕ and ψ can only differ by a constant. Indeed, let X be the Orlicz space $X = L^\Phi((t_1,t_2), dx/(t_2-t_1))$, $0 < t_1 < t_2 < \infty$. Let $X(0)$ be as before the subspace of X of functions having expectation equal to 0. As λf is a C_{t_2}-atom and C^{t_1}-atom simultaneoulsly, for some λ, then

$$\int_{t_1}^{t_2} f\phi = \int_{t_1}^{t_2} f\psi,$$

for all functions in $X(0)$. Given any function in X, $f - Ef \in X(0)$ and repeating again the same arguments we arrive at $\phi - \psi = C$, and this is true for all pairs $0 < t_1 < t_2 < \infty$. Thus, the function $\phi - \psi$ is constant on $(0,\infty)$. Consequently, for all C-atoms a, $T(a) = \int_0^\infty a\phi$.

We conclude the proof showing that $\phi \in \mathrm{BCO}_2$.

Fix $t > 0$ and let f be a fixed function in X_t with $\|f\|_{X_t} \le 1$. Then, since the function $f - E_tf$ is a multiple of a C_t-atom, we have

$$|E_t((\phi - E_t\phi)f)| = |E_t(\phi(f - E_tf))|$$

$$\begin{aligned} |E_t(\phi(f - E_tf))| &= |t^{-1}T(f - E_tf)| \\ &\le t^{-1}\|T\|\|f - E_t\|_{CH^1} \le \|T\|\|f - E_t\|_{X_t} \le 2\|T\|. \end{aligned}$$

Therefore, $\|\phi - E_t\phi\|_{Y_t} \le 4\|T\|$ and

$$E_t\Psi\left(\frac{1}{4\|T\|}|\phi - E_t\phi|\right) \le 1.$$

In a similar way, if $f \in X^t$, the function $\frac{t}{x^2}(f - E^tf)$ is a multiple of a C^t-atom and

$$\|(f - E^tf)\frac{t}{x^2}\|_{CH_1} \le \|f - E^tf\|_{X^t}.$$

Thus

$$\begin{aligned} |E^t((\phi - E^t\phi)f)| &= |E^t(\phi(f - E^tf))| \\ &= |T((f - E^tf)\frac{t}{x^2}| \le \|T\|\|(f - E^tf)\frac{t}{x^2}\|_{CH_1} \\ &\le \|T\|\|f - E^tf\|_{X^t} \le 2\|T\| \end{aligned}$$

and we arrive at

$$E^t\Psi\left(\frac{1}{4\|T\|}|\phi - E^t\phi|\right) \le 1$$

for all $t > 0$. Hence $\|\phi\|_{\mathrm{BCO}_2} \le 2\|T\|$ and the result follows. $\square$

References

[AM] M. A. Ariño and B. Muckenhoupt, *Maximal Functions on Classical Lorentz Spaces and Hardy's Inequality with Weights for Nonincreasing Functions*, Trans. Amer. Math. Soc. **320 (2)** (1990), 727-735.

[AM1] J. Alvarez and M. Milman, *Spaces of Carleson measures, duality and interpolation*, Arkiv för Mat. **25** (1987), 155-173.

[AM2] J. Alvarez and M. Milman, *Interpolation of tent spaces and applications*, Lecture Notes in Math (Function spaces and Applications, Edited by M. Cwikel, J. Peetre, Y. Sagher and H. Wallin) **1302** (1988), 11-21.

[BMR] J. Bastero, M. Milman and F. Ruiz, *Reverse Hölder inequalities and interpolation*, Israel Math. Conf. Proc. **13** (1999), 11-23.

[BR] J. Bastero and F. Ruiz, *Elementary reverse Hölder type inequalities with application to operator interpolation theory*, Proc. Amer. Math. Soc. **124** (1996), 3183-3192.

[BeRu] C. Bennett and K. Rudnick, *On Lorentz Zygmund spaces*, Diss. Math. **CLXXV** (1980).

[BS] C. Bennett and R. Sharpley, *Interpolation of operators*, Academic Press, 1988.

[BL] J. Bergh and J. Löfström, *Interpolation Spaces. An Introduction*, Springer-Verlag, New York, 1976.

[BD] A. Beurling and J. Deny, *Dirichlet spaces*, Proc. Nat. Acad. Sci. **45** (1959), 208-215.

[BK] Y. Brundyi and N. Krugljak, *Interpolation functors and interpolation spaces*, North-Holland, 1991.

[Bl] S. Bloom, *Solving weighted inequalities using the Rubio de Francia algorithm*, Proc. Amer. Math. Soc. **101** (1987), 306-312.

[Bu] S. M. Buckley, *Estimates for operator norms on weighted spaces and reverse Jensen inequalities*, Trans. Amer. Math. Soc. **340** (1993), 253–272.

[Bui] Q. Bui, *Weighted Besov and Triebel spaces: Interpolation by the real method*, Hiroshima Math. J. **3** (1982), 581–605.

[CCS] M. J. Carro, J. Cerdá and J. Soria, *Commutators and interpolation methods*, Ark. Mat. **33** (1995), 199-216.

[CCS1] M. J. Carro, J. Cerdá and J. Soria, *Commutators, interpolation and vector function spaces*, Israel Math. Conf. Proc. **13** (1999), 24-31.

[CCMS] M. J. Carro, J. Cerdá, M. Milman and J. Soria, *Schechter methods of interpolation and commutators*, Math. Nachr. **174** (1995), 35-53.

[Ch] F. Chiarenza, *Regularity for solutions of quasilinear elliptic equations under minimal assumptions*, Potential Analysis **4** (1995), 325-334.

[CLMS] R. Coifman, P. L. Lions, Y. Meyer and S. Semmes, *Compensated compactness and Hardy spaces*, Jour. Math. Pures et Appl. **72** (1993), 247-286.

[CMS] R. Coifman, Y. Meyer and E. Stein, *Some new function spaces and their applications to harmonic analysis*, J. Funct. Anal. **62** (1985), 304-335.

[CRW] R. Coifman, R. Rochberg, and G. Weiss, *Factorization theorems for Hardy spaces in several variables*, Ann. Math. **103** (1976), 611-635.

[Cw] M. Cwikel, *Monotonicity properties of interpolation spaces II*, Arkiv för Mat. **19 (1)** (1981), 123-136.

[Cw1] M. Cwikel, *On* $(L^{p_0}(A_0), L^{p_1}(A_1))$, Proc. Amer. Math. Soc. **44** (1974), 286-292.

[CJM] M. Cwikel, B. Jawerth and M. Milman, *The domain spaces of quasilogarithmic operators*, Trans. Amer. Math. Soc. **317** (1990), 599-609.

[CJMR] M. Cwikel, B. Jawerth, M. Milman and R. Rochberg, *Differential estimates and commutators in interpolation theory.*, Analysis at Urbana vol II, London Math. Soc. Lecture Note Ser. 138, Cambridge Univ. Press, Cambridge-New York, 1989..

[CKMR] M. Cwikel, N. Kalton, M. Milman and R. Rochberg, *A unified approach to derivation mappings* ω *for a class of interpolation methods, (preprint)* (1996).

[DO] V. Dimitriev and V. I. Ovcinnikov, *On interpolation in real method spaces*, Soviet Math. Dokl. **20** (1979), 538-542.

[EM] L. Evans and S. Muller, *Hardy spaces and the two-dimensional Euler equations with nonnegative vorticity*, J. Amer. Math. Soc. **7** (1994), 199–219.

[EOP] W. D. Evans, B. Opic and L. Pick, *Real interpolation with logarithmic functors*, preprint (1996).

[FK] Fan Ming and S. Kaijser, *Complex interpolation with derivatives of analytic functions*, J. Funct. Anal. **120** (1994), 380-402.

[GC] J. García Cuerva, *General endpoints results in extrapolation*, Lecture Notes in Pure and Applied Mathematics (Analysis and Partial Differential Equations) **122** (1990), 161-169.

[GR] J. García Cuerva and J. Rubio de Francia, *Weigthed norm inequalities and related topics*, North-Holland, 1985.

[G] J. Gustavson, *A functional parameter in connection with interpolation of Banach spaces*, Math. Scand. **42 (2)** (1978), 289-305.

[HP] T. Holmstedt and J. Peetre, *On certain functionals arising in the theory of interpolation spaces*, J. Funct. Anal. **4** (1968), 88-94.

[HS] E. Hernández and J. Soria, *Spaces of Lorentz type and complex interpolation*, Arkiv för Mat. **29 (2)** (1991), 203-220.

[IS] T. Iwaniec and C. Sbordone, *Weak minima of variational integrals*, J. Reine Angew. Math. **454** (1994), 143-161.

[JM] B. Jawerth and M. Milman, *Extrapolation theory with applications*, Memoirs Amer. Math. Soc. **440** (1991).

[JM1] B. Jawerth and M. Milman, *Interpolation of weak type spaces*, Math. Z. **201** (1989), 509-519.

[JRW] B. Jawerth, R. Rochberg and G. Weiss, *Commutator and other second order estimates in real interpolation theory*, Arkiv för Mat. **24** (1986), 191-219.

[Ka] N. Kalton, *Nonlinear commutators in interpolation theory*, Memoirs Amer. Math. Soc., Providence, RI. **373** (1988).

[Ka1] N. Kalton, *Differential of complex interpolation precesses for Kothe function spaces.*, Trans. Amer. Math. Soc. **333** (1992), 479–529.

[Ka2] N. Kalton, *Trace class operators and commutators*, J. Funct. Anal. **86** (1989), 41-74.

[KPR] N. Kalton, N. T. Peck, and J. W. Roberts, *An F-space sampler*, London Math. Soc. Lecture Notes 89, Cambridge Univ. Press, Cambridge, 1985.

[K] T. F. Kalugina, *Interpolation of Banach spaces with a functional parameter. The reiteration theorem*, Vestni Moskov Univ., Ser. I Mat. Mec. **30 (6)** (1975), 68-77.

[Kr] P. Krée, *Interpolation d'espaces vectoriels qui ne sont ni normés, ni complets. Applications*, Ann. Inst. Fourier (Grenoble) **17** (1967), 137-174.

[Ku] A. Kufner, O. John and S. Fucik, *Function spaces*, Noordhooff Int. Pub., 1977.

[Li] C. Li, *Compensated compactness, commutators and Hardy spaces. Macquaire University, preprint* (1994).

[LMZ] C. Li, A. McIntosh and K. Zhang, *Higher integrability and reverse Holder inequalities. Macquaire University, preprint* (1993).

[LZ] C. Li, and K. Zhang, *Higher integrability of certain bilinear forms on Orlicz spaces. Macquaire University, preprint* (1994).

[Lu] S. Lu, *Four lectures on real H^p spaces*, Word Scientific, 1995.

[Ma] V. G. Maz'ja, *Sobolev spaces*, Springer-Verlag. Springer Series in Soviet Mathematics, 1985.

[Mi] M. Milman, *Extrapolation and optimal decompositions, with applications to analysis, Lecture Notes in Mathematics 1580*, Springer-Verlag, New York, 1994.

[Mi1] M. Milman, *Integrability of the Jacobian of orientation preserving maps*, Comptes Rend. Acad. Sci. Paris **317** (1993), 539-543.

[Mi3] M. Milman, *Higher order commutators in the real method of interpolation*, Journal D'Analyse Math. **66** (1995), 37-56.

[MR] M. Milman and R. Rochberg, *The role of Cancellation in Interpolation Theory*, Contemporary Math. **189** (1995), 403-419.

[MS] M. Milman and Y. Sagher, *An interpolation theorem*, Arkiv för Mat. **22** (1984), 33-38.

[Mu1] B. Muckenhoupt, *Weighted norm inequalities for the Hardy maximal function*, Trans. Amer. Math. Soc. **165** (1972), 207-226.

[Mu2] B, Muckenhoupt, *Hardy's inequalities with weights*, Studia Math. **44** (1972), 31-38.

[Mu] S. Müller, *Higher integrability of determinants and weak convergence in L^1*, J. Reine Angew Math. **412** (1990), 20-34.

[N1] C. J. Neugebauer, *Weighted norm inequalities for averaging operators of monotone functions*, Publications Matematiques **35** (1991), 429-447.

[N2] C. J. Neugebauer, *Some classical operators on Lorentz space*, Forum Math. **4** (1992), 135-146.

[Pe] C. Perez, *Endpoint estimates for commutators of singular integral operators*, J. Funct. Anal. **128** (1995), 163-185.

[Ro] S. Rolewicz, *Metric linear spaces. MM 56*, PWN, Warsaw, 1972.

[R] R. Rochberg, *Higher order estimates in complex interpolation theory*, Pacific Journal of Math. **174** (1996), 247-267.

[R1] R. Rochberg, *Higher order Hankel forms and commutators in Holomorphic spaces (Berkeley, CA, 1995)*, Math. Sci. Res. Inst. Publ. **33** (1998.), Cambridge Univ. Press, Cambridge, 155-178,.

[RW] R. Rochberg and G. Weiss, *Derivatives of analytic families of Banach spaces*, Ann. Math. **118** (1983), 315-347.

[Sg] Y. Sagher, *Real interpolation with weights*, Indiana Math. J. **30** (1981), 113-121.

[Sw] E. Sawyer, *Boundedness of Classical Operators on Classical Lorentz Spaces*, Studia Math. **96** (1990), 145-158.

[SeT] C. Segovia and J. L. Torrea, *Weighted inequalities for commutators of fractional and singular integrals*, Publ. Math. **35** (1991), 209-235.

[Se] S. Semmes, *A primer on Hardy spaces, and some remarks on a theorem of Evans and Müller*, Comm in Part Diff. Eq. **19** (1994), 273-319.

[So] J. Soria, *Weighted Tent spaces*, Math. Nach. **155** (1992), 231-256.

[We] T. Weidl, *Cwikel type estimates in non-power ideals*, Math. Narich. **176** (1995), 315-334.

[Wo] T. H. Wolff, *A note on interpolation spaces*, Lecture Notes in Mathematics **908** (1982), Springer Verlag, 199-204.

[Wu] S. Wu, *A wavelet characterization for weighted Hardy spaces*, Revista Mat. Iber. **8** (1992), 329-349.

[Zy] A. Zygmund, *Trigonometric series*, Cambridge Univ. Press, New York, 1959.

Editorial Information

To be published in the *Memoirs*, a paper must be correct, new, nontrivial, and significant. Further, it must be well written and of interest to a substantial number of mathematicians. Piecemeal results, such as an inconclusive step toward an unproved major theorem or a minor variation on a known result, are in general not acceptable for publication. Papers appearing in *Memoirs* are generally longer than those appearing in *Transactions*, which shares the same editorial committee.

As of May 31, 2001, the backlog for this journal was approximately 7 volumes. This estimate is the result of dividing the number of manuscripts for this journal in the Providence office that have not yet gone to the printer on the above date by the average number of monographs per volume over the previous twelve months, reduced by the number of volumes published in four months (the time necessary for preparing a volume for the printer). (There are 6 volumes per year, each containing at least 4 numbers.)

A Consent to Publish and Copyright Agreement is required before a paper will be published in the *Memoirs*. After a paper is accepted for publication, the Providence office will send a Consent to Publish and Copyright Agreement to all authors of the paper. By submitting a paper to the *Memoirs*, authors certify that the results have not been submitted to nor are they under consideration for publication by another journal, conference proceedings, or similar publication.

Information for Authors

Memoirs are printed from camera copy fully prepared by the author. This means that the finished book will look exactly like the copy submitted.

The paper must contain a *descriptive title* and an *abstract* that summarizes the article in language suitable for workers in the general field (algebra, analysis, etc.). The *descriptive title* should be short, but informative; useless or vague phrases such as "some remarks about" or "concerning" should be avoided. The *abstract* should be at least one complete sentence, and at most 300 words. Included with the footnotes to the paper should be the 2000 *Mathematics Subject Classification* representing the primary and secondary subjects of the article. The classifications are accessible from `www.ams.org/msc/`. The list of classifications is also available in print starting with the 1999 annual index of *Mathematical Reviews*. The Mathematics Subject Classification footnote may be followed by a list of *key words and phrases* describing the subject matter of the article and taken from it. Journal abbreviations used in bibliographies are listed in the latest *Mathematical Reviews* annual index. The series abbreviations are also accessible from `www.ams.org/publications/`. To help in preparing and verifying references, the AMS offers MR Lookup, a Reference Tool for Linking, at `www.ams.org/mrlookup/`. When the manuscript is submitted, authors should supply the editor with electronic addresses if available. These will be printed after the postal address at the end of the article.

Electronically prepared manuscripts. The AMS encourages electronically prepared manuscripts, with a strong preference for $\mathcal{A}\mathcal{M}\mathcal{S}$-LaTeX. To this end, the Society has prepared $\mathcal{A}\mathcal{M}\mathcal{S}$-LaTeX author packages for each AMS publication. Author packages include instructions for preparing electronic manuscripts, the *AMS Author Handbook*, samples, and a style file that generates the particular design specifications of that publication series. Though $\mathcal{A}\mathcal{M}\mathcal{S}$-LaTeX is the highly preferred format of TeX, author packages are also available in $\mathcal{A}\mathcal{M}\mathcal{S}$-TeX.

Authors may retrieve an author package from e-MATH starting from `www.ams.org/tex/` or via FTP to `ftp.ams.org` (login as `anonymous`, enter username as password, and type `cd pub/author-info`). The *AMS Author Handbook* and the *Instruction Manual* are available in PDF format following the author packages link from `www.ams.org/tex/`. The author package can be obtained free of charge by sending email to `pub@ams.org` (Internet) or from the Publication Division, American Mathematical Society, P.O. Box 6248, Providence, RI 02940-6248. When requesting an author package, please specify $\mathcal{A}\mathcal{M}\mathcal{S}$-LaTeX or $\mathcal{A}\mathcal{M}\mathcal{S}$-TeX, Macintosh or IBM (3.5) format, and the publication in which your paper will appear. Please be sure to include your complete mailing address.

Sending electronic files. After acceptance, the source file(s) should be sent to the Providence office (this includes any TeX source file, any graphics files, and the DVI or PostScript file).

Before sending the source file, be sure you have proofread your paper carefully. The files you send must be the EXACT files used to generate the proof copy that was accepted for publication. For all publications, authors are required to send a printed copy of their paper, which exactly matches the copy approved for publication, along with any graphics that will appear in the paper.

TeX files may be submitted by email, FTP, or on diskette. The DVI file(s) and PostScript files should be submitted only by FTP or on diskette unless they are encoded properly to submit through email. (DVI files are binary and PostScript files tend to be very large.)

Electronically prepared manuscripts can be sent via email to `pub-submit@ams.org` (Internet). The subject line of the message should include the publication code to identify it as a Memoir. TeX source files, DVI files, and PostScript files can be transferred over the Internet by FTP to the Internet node `e-math.ams.org` (130.44.1.100).

Electronic graphics. Comprehensive instructions on preparing graphics are available at `www.ams.org/jourhtml/graphics.html`. A few of the major requirements are given here.

Submit files for graphics as EPS (Encapsulated PostScript) files. This includes graphics originated via a graphics application as well as scanned photographs or other computer-generated images. If this is not possible, TIFF files are acceptable as long as they can be opened in Adobe Photoshop or Illustrator. No matter what method was used to produce the graphic, it is necessary to provide a paper copy to the AMS.

Authors using graphics packages for the creation of electronic art should also avoid the use of any lines thinner than 0.5 points in width. Many graphics packages allow the user to specify a "hairline" for a very thin line. Hairlines often look acceptable when proofed on a typical laser printer. However, when produced on a high-resolution laser imagesetter, hairlines become nearly invisible and will be lost entirely in the final printing process.

Screens should be set to values between 15% and 85%. Screens which fall outside of this range are too light or too dark to print correctly. Variations of screens within a graphic should be no less than 10%.

Inquiries. Any inquiries concerning a paper that has been accepted for publication should be sent directly to the Electronic Prepress Department, American Mathematical Society, P. O. Box 6248, Providence, RI 02940-6248.

Editors

This journal is designed particularly for long research papers, normally at least 80 pages in length, and groups of cognate papers in pure and applied mathematics. Papers intended for publication in the *Memoirs* should be addressed to one of the following editors. In principle the Memoirs welcomes electronic submissions, and some of the editors, those whose names appear below with an asterisk (*), have indicated that they prefer them. However, editors reserve the right to request hard copies after papers have been submitted electronically. Authors are advised to make preliminary email inquiries to editors about whether they are likely to be able to handle submissions in a particular electronic form.

Algebra to CHARLES CURTIS, Department of Mathematics, University of Oregon, Eugene, OR 97403-1222 email: `cwc@darkwing.uoregon.edu`

Algebraic geometry and commutative algebra to LAWRENCE EIN, Department of Mathematics, University of Illinois, 851 S. Morgan (M/C 249), Chicago, IL 60607-7045; email: `ein@uic.edu`

Algebraic topology and cohomology of groups to STEWART PRIDDY, Department of Mathematics, Northwestern University, 2033 Sheridan Road, Evanston, IL 60208-2730; email: `priddy@math.nwu.edu`

Combinatorics and Lie theory to SERGEY FOMIN, Department of Mathematics, University of Michigan, Ann Arbor, Michigan 48109-1109; email: `fomin@math.lsa.umich.edu`

Complex analysis and complex geometry to DUONG H. PHONG, Department of Mathematics, Columbia University, 2990 Broadway, New York, NY 10027-0029; email: `phong@math.columbia.edu`

***Differential geometry and global analysis** to LISA C. JEFFREY, Department of Mathematics, University of Toronto, 100 St. George St., Toronto, ON Canada M5S 3G3; email: `jeffrey@math.toronto.edu`

***Dynamical systems and ergodic theory** to ROBERT F. WILLIAMS, Department of Mathematics, University of Texas, Austin, Texas 78712-1082; email: `bob@math.utexas.edu`

Functional analysis and operator algebras to BRUCE E. BLACKADAR, Department of Mathematics, University of Nevada, Reno, NV 89557; email: `bruceb@math.unr.edu`

Geometric topology, knot theory and hyperbolic geometry to ABIGAIL A. THOMPSON, Department of Mathematics, University of California, Davis, Davis, CA 95616-5224; email: `thompson@math.ucdavis.edu`

Harmonic analysis, representation theory, and Lie theory to ROBERT J. STANTON, Department of Mathematics, The Ohio State University, 231 West 18th Avenue, Columbus, OH 43210-1174; email: `stanton@math.ohio-state.edu`

***Logic** to THEODORE SLAMAN, Department of Mathematics, University of California, Berkeley, CA 94720-3840; email: `slaman@math.berkeley.edu`

Number theory to MICHAEL J. LARSEN, Department of Mathematics, Indiana University, Bloomington, IN 47405; email: `larsen@math.indiana.edu`

***Ordinary differential equations, partial differential equations, and applied mathematics** to PETER W. BATES, Department of Mathematics, Brigham Young University, 292 TMCB, Provo, UT 84602-1001; email: `peter@math.byu.edu`

***Partial differential equations and applied mathematics** to BARBARA LEE KEYFITZ, Department of Mathematics, University of Houston, 4800 Calhoun Road, Houston, TX 77204-3476; email: `keyfitz@uh.edu`

***Probability and statistics** to KRZYSZTOF BURDZY, Department of Mathematics, University of Washington, Box 354350, Seattle, Washington 98195-4350; email: `burdzy@math.washington.edu`

***Real and harmonic analysis and geometric partial differential equations** to WILLIAM BECKNER, Department of Mathematics, University of Texas, Austin, TX 78712-1082; email: `beckner@math.utexas.edu`

All other communications to the editors should be addressed to the Managing Editor, WILLIAM BECKNER, Department of Mathematics, University of Texas, Austin, TX 78712-1082; email: `beckner@math.utexas.edu`.

Selected Titles in This Series

(Continued from the front of this publication)

700 **Vicente Cortés,** A new construction of homogeneous quaternionic manifolds and related geometric structures, 2000

699 **Alexander Fel′shtyn,** Dynamical zeta functions, Nielsen theory and Reidemeister torsion, 2000

698 **Andrew R. Kustin,** Complexes associated to two vectors and a rectangular matrix, 2000

697 **Deguang Han and David R. Larson,** Frames, bases and group representations, 2000

696 **Donald J. Estep, Mats G. Larson, and Roy D. Williams,** Estimating the error of numerical solutions of systems of reaction-diffusion equations, 2000

695 **Vitaly Bergelson and Randall McCutcheon,** An ergodic IP polynomial Szemerédi theorem, 2000

694 **Alberto Bressan, Graziano Crasta, and Benedetto Piccoli,** Well-posedness of the Cauchy problem for $n \times n$ systems of conservation laws, 2000

693 **Doug Pickrell,** Invariant measures for unitary groups associated to Kac-Moody Lie algebras, 2000

692 **Mara D. Neusel,** Inverse invariant theory and Steenrod operations, 2000

691 **Bruce Hughes and Stratos Prassidis,** Control and relaxation over the circle, 2000

690 **Robert Rumely, Chi Fong Lau, and Robert Varley,** Existence of the sectional capacity, 2000

689 **M. A. Dickmann and F. Miraglia,** Special groups: Boolean-theoretic methods in the theory of quadratic forms, 2000

688 **Piotr Hajłasz and Pekka Koskela,** Sobolev met Poincaré, 2000

687 **Guy David and Stephen Semmes,** Uniform rectifiability and quasiminimizing sets of arbitrary codimension, 2000

686 **L. Gaunce Lewis, Jr.,** Splitting theorems for certain equivariant spectra, 2000

685 **Jean-Luc Joly, Guy Metivier, and Jeffrey Rauch,** Caustics for dissipative semilinear oscillations, 2000

684 **Harvey I. Blau, Bangteng Xu, Z. Arad, E. Fisman, V. Miloslavsky, and M. Muzychuk,** Homogeneous integral table algebras of degree three: A trilogy, 2000

683 **Serge Bouc,** Non-additive exact functors and tensor induction for Mackey functors, 2000

682 **Martin Majewski,** ational homotopical models and uniqueness, 2000

681 **David P. Blecher, Paul S. Muhly, and Vern I. Paulsen,** Categories of operator modules (Morita equivalence and projective modules, 2000

680 **Joachim Zacharias,** Continuous tensor products and Arveson's spectral C^*-algebras, 2000

679 **Y. A. Abramovich and A. K. Kitover,** Inverses of disjointness preserving operators, 2000

678 **Wilhelm Stannat,** The theory of generalized Dirichlet forms and its applications in analysis and stochastics, 1999

677 **Volodymyr V. Lyubashenko,** Squared Hopf algebras, 1999

676 **S. Strelitz,** Asymptotics for solutions of linear differential equations having turning points with applications, 1999

675 **Michael B. Marcus and Jay Rosen,** Renormalized self-intersection local times and Wick power chaos processes, 1999

674 **R. Lawther and D. M. Testerman,** A_1 subgroups of exceptional algebraic groups, 1999

673 **John Lott,** Diffeomorphisms and noncommutative analytic torsion, 1999

672 **Yael Karshon,** Periodic Hamiltonian flows on four dimensional manifolds, 1999

For a complete list of titles in this series, visit the AMS Bookstore at **www.ams.org/bookstore/**.